SHOP MANUALS

JOHN DEERE

SHOP MANUAL

Information and Instructions

This shop manual contains several sections each covering a specific group of wheel type tractors. The Tab Index on the preceding page can be used to locate the section pertaining to each group of tractors. Each section contains the necessary specifications and the brief but terse procedural data needed by a mechanic when repairing a tractor on which he has had no previous actual experience.

Within each section, the material is arranged in a systematic order beginning with an index which is followed immediately by a Table of Condensed Service Specifications. These specifications include dimensions, fits, clearances and timing instructions. Next in order of arrangement is the procedures paragraphs.

In the procedures paragraphs, the order of presentation starts with the front axle system and steering and proceeding toward the rear axle. The last paragraphs are devoted to the power take-off and power lift systems. Interspersed where needed are additional tabular specifications pertaining to wear limits, torquing, etc.

HOW TO USE THE INDEX

Suppose you want to know the procedure for R&R (remove and reinstall) of the engine camshaft. Your first step is to look in the index under the main heading of ENGINE until you find the entry "Camshaft." Now read to the right where under the column covering the tractor you are repairing, you will find a number which indicates the beginning paragraph pertaining to the camshaft. To locate this wanted paragraph in the manual, turn the pages until the running index appearing on the top outside corner of each page contains the number you are seeking. In this paragraph you will find the information concerning the removal of the camshaft.

More information available at Clymer.com
Phone: 805-498-6703

Haynes Publishing Group
Sparkford Nr Yeovil
Somerset BA22 7JJ England

Haynes North America, Inc
859 Lawrence Drive
Newbury Park
California 91320 USA

ISBN-10: 0-87288-501-1
ISBN-13: 978-0-87288-501-1

© **Haynes North America, Inc. 1992**
With permission from J.H. Haynes & Co. Ltd.

Clymer is a registered trademark of Haynes North America, Inc.

Printed in Malaysia

Cover art by Sean Keenan

Common spark plug conditions

NORMAL
Symptoms: Brown to grayish-tan color and slight electrode wear. Correct heat range for engine and operating conditions.
Recommendation: When new spark plugs are installed, replace with plugs of the same heat range.

WORN
Symptoms: Rounded electrodes with a small amount of deposits on the firing end. Normal color. Causes hard starting in damp or cold weather and poor fuel economy.
Recommendation: Plugs have been left in the engine too long. Replace with new plugs of the same heat range. Follow the recommended maintenance schedule.

CARBON DEPOSITS
Symptoms: Dry sooty deposits indicate a rich mixture or weak ignition. Causes misfiring, hard starting and hesitation.
Recommendation: Make sure the plug has the correct heat range. Check for a clogged air filter or problem in the fuel system or engine management system. Also check for ignition system problems.

ASH DEPOSITS
Symptoms: Light brown deposits encrusted on the side or center electrodes or both. Derived from oil and/or fuel additives. Excessive amounts may mask the spark, causing misfiring and hesitation during acceleration.
Recommendation: If excessive deposits accumulate over a short time or low mileage, install new valve guide seals to prevent seepage of oil into the combustion chambers. Also try changing gasoline brands.

OIL DEPOSITS
Symptoms: Oily coating caused by poor oil control. Oil is leaking past worn valve guides or piston rings into the combustion chamber. Causes hard starting, misfiring and hesitation.
Recommendation: Correct the mechanical condition with necessary repairs and install new plugs.

GAP BRIDGING
Symptoms: Combustion deposits lodge between the electrodes. Heavy deposits accumulate and bridge the electrode gap. The plug ceases to fire, resulting in a dead cylinder.
Recommendation: Locate the faulty plug and remove the deposits from between the electrodes.

TOO HOT
Symptoms: Blistered, white insulator, eroded electrode and absence of deposits. Results in shortened plug life.
Recommendation: Check for the correct plug heat range, over-advanced ignition timing, lean fuel mixture, intake manifold vacuum leaks, sticking valves and insufficient engine cooling.

PREIGNITION
Symptoms: Melted electrodes. Insulators are white, but may be dirty due to misfiring or flying debris in the combustion chamber. Can lead to engine damage.
Recommendation: Check for the correct plug heat range, over-advanced ignition timing, lean fuel mixture, insufficient engine cooling and lack of lubrication.

HIGH SPEED GLAZING
Symptoms: Insulator has yellowish, glazed appearance. Indicates that combustion chamber temperatures have risen suddenly during hard acceleration. Normal deposits melt to form a conductive coating. Causes misfiring at high speeds.
Recommendation: Install new plugs. Consider using a colder plug if driving habits warrant.

DETONATION
Symptoms: Insulators may be cracked or chipped. Improper gap setting techniques can also result in a fractured insulator tip. Can lead to piston damage.
Recommendation: Make sure the fuel anti-knock values meet engine requirements. Use care when setting the gaps on new plugs. Avoid lugging the engine.

MECHANICAL DAMAGE
Symptoms: May be caused by a foreign object in the combustion chamber or the piston striking an incorrect reach (too long) plug. Causes a dead cylinder and could result in piston damage.
Recommendation: Repair the mechanical damage. Remove the foreign object from the engine and/or install the correct reach plug.

SHOP MANUAL
JOHN DEERE

SERIES
2750—2755—2855N—2955

Tractor serial number is stamped in right side of front support and is also located on a plate attached to right side of front support. Engine serial number is stamped on a plate on right side of engine cylinder block.

INDEX (By Starting Paragraph)

INDEX (Cont.)

DUAL DIMENSIONS

This shop manual provides specifications in both Metric (SI) and U.S. Customary measurement systems. The first specification is given in the measuring system used during manufacture; the second specification (given in parentheses) is the converted measurement. For instance, the specification "0.28 mm (0.011 inch)" indicates that the equipment was manufactured using the metric measurement system and the U.S. equivalent of 0.28 mm is 0.011 inch.

CONDENSED SERVICE DATA

	2750 CS	2750 TSS	2755 CS	2755 TSS	2855N	2955 TSS
GENERAL						
Engine Model...........	4239TL-04	4239TL-05	CD4239TL005	4239TL009	4239TL008	CD6359DL009
No. of Cylinders	4	4	4	4	4	6
Bore			106.5 mm (4.19 in.)			
Stroke.............			110 mm (4.33 in.)			
Displacement..........	3.92 L (239 cid.)	3.92 L (239 cid.)	3.92 L (239 cid.)	3.92 L (429 cid.)	3.92 L (239 cid.)	5.88 L (359 cid.)
Compression Ratio.......			17.8:1			
Battery Terminal Grounded.............			Negative			
TUNE-UP						
Firing Order	1-3-4-2	1-3-4-2	1-3-4-2	1-3-4-2	1-3-4-2	1-5-3-6-2-4
Valve Clearance, Inlet			0.35 mm (0.014 in.)			
Exhaust			0.45 mm (0.018 in.)			
Engine Low Idle—Rpm...	700-800	700-800	750-850	750-850	750-850	750-850
Engine High Idle—Rpm ..	2610-2660	2610-2660	2610-2660	2410-2510	2410-2510	2410-2510
Pto Power at Rpm	*1	*1	*2	*3	*4	*5

*1 — 2750 - 56 kW (75 Hp) @ 2500 rpm.

*2 — 2755 with Collar Shift Transmission - 56 kW (75 Hp) @ 2500 rpm.

*3 — 2755 with Synchronized transmission - 56 kW (75 Hp) @ 2300 rpm.

*4 — 2855N - 60 kW (80 Hp) @ 2300 rpm.

*5 — 2955 - 63 kW (85 Hp) @ 2300 rpm.

SIZES—CLEARANCES

Crankshaft Main Journal Diameter	79.34-79.36 mm (3.123-3.124 in.)
Crankpin Diameter.......	69.80-69.82 mm (2.748-2.749 in.)
Piston Pin Diameter— Small pin	34.92-34.93 mm (1.374-1.375 in.)
Large pin	41.27-41.28 mm (1.624-1.625 in.)
Main Bearing Clearance ..	0.03-0.10 mm (0.0012-0.004 in.)
Rod Bearing Clearance ...	0.03-0.10 mm (0.0012-0.004 in.)
Camshaft Journal Clearance	0.08-0.13 mm (0.003-0.005 in.)
Crankshaft End Play— With 2-Piece Thrust Bearing	0.05-0.20 mm (0.002-0.008 in.)

CONDENSED SERVICE DATA (CONT.)

	2750 CS	2750 TSS	2755 CS	2755 TSS	2855N	2955 TSS

SIZES—CLEARANCES (Cont.)

Crankshaft End Play (Cont.)
  With 5-Piece Thrust
    Bearing . . . .   0.025-0.43 mm
        (0.001-0.017 in.)

  All Dubuque Built
    Engines *** . . . .   0.025-0.33 mm
        (0.001-0.013 in.)

Camshaft End Play . . . .   0.05-0.23 mm
        (0.002-0.009 in.)

Piston Skirt Clearance—
  All Naturally Aspirated
    Engines . . . .   0.08-0.14 mm
        (0.003-0.005 in.)

  Turbocharged Engines,
  Dubuque Built
    Engines *** . . . .   0.08-0.15 mm
        (0.003-0.006 in.)

  Engines Built At Saran, France
    Before CD600000 . . . .   0.14-0.20 mm
        (0.005-0.008 in.)
    After CD599999 . . . . .   0.08-0.15 mm
        (0.003-0.006 in.)

*** First 2 characters of engine Serial Number will be "TO".

	2750 CS	2750 TSS	2755 CS	2755 TSS	2855N TSS	2955 TSS

CAPACITIES

All capacities are approximate and may be different than shown.

Cooling System
  With Sound Gard Body .
  w/o Sound Gard Body . .

Item	2750 CS	2750 TSS	2755 CS	2755 TSS	2855N	2955 TSS
Cooling System — With Sound Gard Body	15.0 L (4 gal.)	15.0 L (4 gal.)	15.0 L (4 gal.)	15.0 L (4 gal.)	15.0 L (4 gal.)	19.0 L (5.0 gal.)
w/o Sound Gard Body	13.0 L (3.4 gal.)	13.0 L (3.4 gal.)	13.0 L (3.4 gal.)	13.0 L (3.4 gal.)	13.0 L (3.4 gal.)	17.0 L (4.5 gal.)
Air Conditioning			1.8 kg (4 lbs.)			
Crankcase (with filter)	8.5 L (9.2 qt.)	8.5 L (9.2 qt.)	10.0 L (10.4 qt.)	10.0 L (10.4 qt.)	10.5 L (11.2 qt.)	11.5 L (12 qt.)
Fuel Tank	98 L (25.9 gal.)	98 L (25.9 gal.)	84 L (22.2 gal.)	84 L (22.2 gal.)	84 L (22.2 gal.)	121 L (32 gal.)
Auxiliary Fuel Tank	....	....	52.0 L (13.7 gal.)	52.0 L (13.7 gal.)	52.0 L (13.7 gal.)	52.0 L (13.7 gal.)
Transmission & Hydraulic— Synchronized	....	56-64 L (14.7-16.9 gal.)	....	50-55 L	50-55 L (13.2-14.5 gal.)	50-55 L
Collar Shift	56-64 L (14.7-16.9 gal.)	....	46 L (12.2 gal.)	....	....	....
Front-Wheel Drive— Axle Housing	....	5.3 L (5.6 qt.)	....	5.3 L (5.6 qt.)	5.3 L (5.6 qt.)	6.5 L (6.9 qt.)
Planetary (each)	....	0.75 L (0.8 qt.)	....	0.75 L (0.8 qt.)	0.75 L (0.8 qt.)	1.0 L (1.04 qt.)

FRONT SYSTEM
(TWO-WHEEL DRIVE)

AXLE AND SUPPORT

All Two-Wheel Drive Models

1. AXLE CENTER MEMBER. To remove front axle assembly, raise front of tractor in such a way that it will not interfere with the removal of the axle. A hoist may be attached to front support or special stands can be attached to the side rails. Remove front wheels and weights, then support the axle with a suitable jack or special safety stand to prevent tipping while permitting the axle to be lowered and moved safely. Disconnect drag link from the bellcrank (13—Figs. 2 or 3) of models without hydrostatic steering. On models with hydrostatic steering, disconnect hoses from the steering cylinders and cover openings to prevent the entry of dirt. On all models, remove the slotted nut (1—Figs. 1, 2, 3, 4 or 5), special pivot bolt (6), washers (2) and shims (3). Move axle rearward until free from pivot tube (4) at front and pin (7) at

rear. Lower the axle assembly and carefully roll axle away from under front support.

On models without hydrostatic steering, a bell-crank (13—Figs. 2 or 3) is supported in tapered roller bearings, which should be lubricated by packing with EP multipurpose grease. The tapered roller bearings for bellcrank should be correctly located vertically by adding the required number of shims (11) between the bearing cup of upper bearing and bore in axle center member. Correct selection of shims is determined by measuring the depth of bearing bore, then

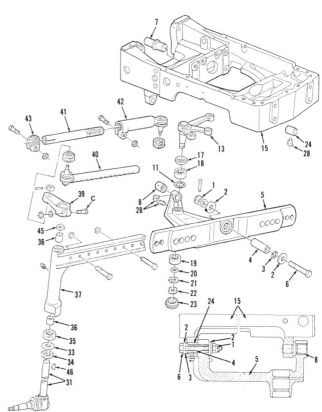

Fig. 2—Adjustable tread standard front axle used on models without hydrostatic steering.

1.	Slotted nut		
2.	Washer	23.	Cap
3.	Shim	24.	Bushing
4.	Pivot tube	28.	Grease fittings
5.	Axle	31.	Spindle
6.	Pivot bolt	33.	Thrust washer
7.	Rear pivot	34.	Lower seal
8.	Bushing	35.	Thrust bearing
11.	Shim	36.	Bushings
13.	Steering bellcrank	37.	Axle extension
15.	Front support	39.	Steering arm
17.	Sealing disc	40.	Tie rod end
18.	Upper bearing	41.	Tube
19.	Lower bearing	42.	Tie rod end
20.	Washer	43.	Clamp
21.	Lock washer	45.	Washer
22.	Nut	46.	Woodruff key

Fig. 1—Fixed tread front axle of type used on 2755 GP model.

1.	Slotted nut	30.	Washer
2.	Washer	31.	Spindle
3.	Shim	32.	Thrust washers
4.	Pivot tube	33.	Thrust washer
5.	Axle	36.	Bushings
6.	Pivot bolt	39.	Steering arm
7.	Rear pivot	40.	Tie rod end
8.	Bushing	41.	Tube
15.	Front support	42.	Tie rod end
24.	Bushing	43.	Clamp
28.	Grease fittings	44.	Clamp

subtracting 18.4 mm (0.724 inch). The result is thickness of shims (11) required. Shim thickness can be rounded to the nearest tenth (0.1) mm when assembling. Tighten the special nut (22) to 60 N·m (45 ft.-lbs.) of torque.

On all models, check bushing (8—Figs. 1, 2, 3, 4 or 5), front pivot tube (4), rear pivot pin (7) and bushing (24) for excessive wear and renew if necessary. Be sure to align passage in bushing with appropriate grease fitting if new bushing is installed. Crossed grooves in bushing (8) should be down and slot in bushing (24) should be up.

Reverse removal procedure when assembling. Tighten slotted nut (1) to 300 N·m (220 ft.-lbs.) torque. End clearance of axle center member should be 0-0.5 mm (0-0.02 inch) and is adjusted by adding or removing shims (3). If cotter pin cannot be installed through slotted nut (1) after tightening to specified torque, tighten enough to align hole with next slot, then install cotter pin.

2. FRONT WHEEL BEARINGS. To remove front wheel hub and bearings, raise and support the front axle extension, then unbolt and remove the tire and wheel assembly. Remove the cap (12—Fig. 6), cotter pin (9), slotted nut (10), washer (8) and outer bearing cone (7). Slide the hub assembly from spindle axle

shaft. Remove and install new collar (2), from models so equipped, if scored or otherwise damaged. Hub is slotted to facilitate removal of bearing cups (4 and 6). Seal (1) used with collar (2) should be driven onto axle with numbered side of seal out toward driving tool. Fill space between seal lips of all seals with EP multipurpose grease and pack bearings liberally with wheel bearing grease. Reassemble by reversing disassembly procedure. Tighten slotted nut (10) to a torque of 50 N·m (35 ft.-lbs.) while rotating hub. Back nut off to nearest slot and install cotter pin (9), then install cap (12).

3. SPINDLES AND BUSHINGS. To remove spindle (31—Figs. 1, 2, 3, 4 or 5), first remove the wheel and hub as outlined in paragraph 2. Disconnect tie rod end from steering arm (9). On models with hydrostatic steering, detach the steering cylinder from the steering arm. On all models, remove cap screw and washers from top of spindle. Remove steering arm (39), shims and washers from top of spindle, then lower spindle out of axle extension. Models with clamping-type steering arm (39—Fig. 2) are not

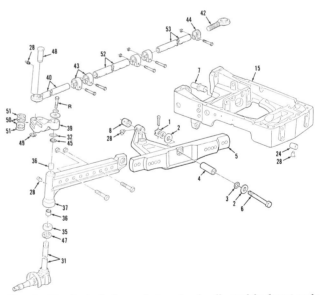

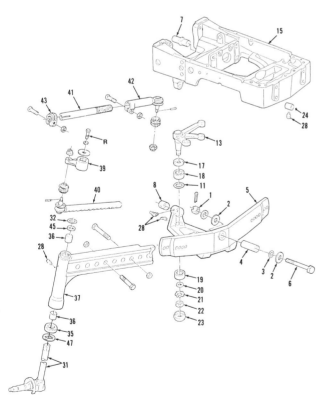

Fig. 3—Exploded view of swept back adjustable front axle used on some models without hydrostatic steering. Refer to Fig. 2 for legend except the following.

32. Washer   47. Washer

Fig. 4—Exploded view of standard adjustable front axle used on models with hydrostatic steering. Steering cylinders are attached to steering arms (39).

1. Slotted nut
2. Washer
3. Shim
4. Pivot tube
5. Axle
6. Pivot bolt
7. Rear pivot
8. Bushing
15. Front support
24. Bushing
28. Grease fittings
31. Spindle
32. Shims
35. Thrust bearing
36. Bushings

37. Axle extension
39. Steering arm
40. Tie rod end
42. Tie rod end
43. Clamps
44. Clamp
45. Washer
47. Washer
48. Pin
49. Snap ring
50. Seals
51. Rings
52. Tube
53. Tube

equipped with shims. On all models, remove thrust bearing, lower seal and washer from spindle. Clean and inspect parts for wear or other damage and renew as necessary.

Install lower bushing (36—Figs. 1, 2, 3, 4 or 5) until flush with bottom of bore. Upper bushing should be installed 8.5 mm (0.33 inch) below top of bore on models with fixed tread axle, 11 mm (0.430 inch) below top of bore for models with heavy-duty adjustable axle. Install upper bushings flush with counterbore of adjustable axles that are not heavy-duty. Bushings are presized and should require no final sizing if carefully installed. Inside diameter of new installed bushings should be 38.11-38.21 mm (1.500-1.504 inches).

When reassembling, install thrust bearing (35—Figs. 2, 3, 4 or 5) on spindle so that numbered side of bearing is facing upward. Add or remove shims (32—Figs. 1, 3, 4 or 5) to adjust spindle end play on models with cap screw retaining steering arm to spindle. On models with steering arm (39—Fig. 2) clamped to spindle, relocate steering arm to adjust end play of spindle. Spindle end play should be 0.76 mm (0.030 inch). Tighten steering arm retaining clamping screw (C) to a torque of 120 N·m (85 ft.-lbs.) for all models so equipped. Tighten the retaining cap screw (R—Figs. 1, 3, 4 or 5) to 230 N·m (170 ft.-lbs.) torque. Balance of reassembly is the reverse of disassembly.

4. TIE RODS AND TOE-IN. Models without hydrostatic steering are equipped with two tie rods extending from left and right steering arms to steering bellcrank (13—Figs. 2 or 3). Models with hydrostatic steering are equipped with one tie rod extending between left and right steering arms. Recommended toe-in is 3-6 mm ($\frac{1}{8}$-$\frac{1}{4}$ inch) for all models. To adjust toe-in of models with tie rod ends as shown in Figs. 1, 2 or 3, loosen or remove bolts in outer clamps (43) of both tie rods and loosen bolts at opposite end clamp (44—Fig. 1) or rod end (42—Figs. 2 or 3). When adjusting models with two tie rods, adjust each side an equal amount to obtain proper toe-in and center steering wheel. Install and tighten clamp bolts. To adjust toe-in of adjustable axle models with hydrostatic steering, loosen bolt in clamp (44—Figs. 4 or 5), remove snap ring (49) and withdraw pin (48). Turn tie rod end (42) in or out of tie rod tube (53) as required to obtain proper toe-in. Connect tie rod to steering arm and tighten clamp bolt.

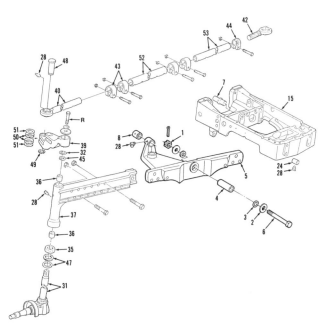

Fig. 5—Exploded view of heavy-duty adjustable front axle used on some models with hydrostatic steering.

1. Slotted nut	28. Grease fittings	
2. Washer	31. Spindle	44. Clamp
3. Shim	32. Shims	45. Washer
4. Pivot tube	35. Thrust bearing	47. Washer
5. Axle	36. Bushings	48. Pin
6. Pivot bolt	37. Axle extension	49. Snap ring
7. Rear pivot	39. Steering arm	50. Seals
8. Bushing	40. Tie rod end	51. Rings
15. Front support	42. Tie rod end	52. Tube
24. Bushing	43. Clamps	53. Tube

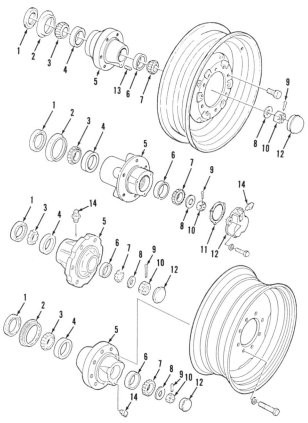

Fig. 6—Exploded view of four types of front wheel hubs used on two-wheel drive models. Seal collar (2) is used for seals that are pressed on spindle hub.

1. Oil seal		8. Washer
2. Collar		9. Cotter pin
3. Bearing cone		10. Slotted nut
4. Bearing cup		11. Gasket
5. Hub		12. Cap
6. Bearing cup		13. Plug
7. Bearing cone		14. Lubrication fitting

FRONT-WHEEL DRIVE SYSTEM
(ALL MODELS SO EQUIPPED)

The mechanical front-wheel drive (MFWD) available on these models uses an APL-300 or APL-700 series front drive axle unit manufactured by Zahnradfabrik Passau GmbH (ZF). There are some differences between the front-wheel drive systems used on these models that will be referred to in the servicing and testing instructions that follow. Please note the following front-wheel drive axle applications:

Tractor Model	Drive System Model
2750 (before SN 583 919 L)............	APL-345
2750 (after SN 583 918 L).............	APL-735
2755	APL-735
2855N	APL-325N
2955	APL-350

The unit is controlled by an electric solenoid/hydraulic valve that operates a multiple disc clutch. This clutch is located in the bottom front of the synchronized transmission case. The gears that drive the clutch are integral with the synchronized transmission. A drive shaft with two "U" joints connects the clutch unit to front axle. Front drive axle may be equipped with a self-locking differential and all models have a planetary drive at each of the front wheel hubs.

> **CAUTION: When servicing a front-wheel drive equipped tractor with rear wheels supported off ground, engine running and transmission in gear, always raise front wheels off ground too. Loss of electrical power or transmission hydraulic system pressure will engage front driving wheels. Rear wheels will be pulled off support if front wheels are not raised.**

TROUBLE-SHOOTING

All Models with MFWD

5. Some problems that may occur with front-wheel drive and their possible causes are as follows:

1. Front-wheel drive not engaging. Could be caused by:
   a. Dash electrical switch defective. Check to be sure that electrical circuit is open when key switch and dash "MFWD" switch are both ON. (The engine does not need to be running when making this check.) Outside of control solenoid should NOT have strong magnetic field as checked by placing a screwdriver near solenoid.
   b. Solenoid/hydraulic valve spool sticking open. The valve may stick if the valve attaching screws are improperly or unevenly tightened. Refer to paragraphs 17 and 18 for tests.
   c. Disc clutch pack worn. To check for slippage with engine stopped, set parking brake, slide the drive shaft shield forward and remove the rear guard mounting bracket. Raise the left front wheel off ground using a floor jack, then use a pipe wrench or other suitable tool attached to the clutch output shaft to measure torque required to cause clutch to slip. Clutch should not slip at less than the following torque:

2750 models	1200 N·m
	(885 ft.-lbs.)
2755 high-clearance tractors..........	1300 N·m
	(960 ft.-lbs.)
2955 high-clearance tractors..........	1300 N·m
	(960 ft.-lbs.)
2955 models with front pto	1300 N·m
	(960 ft.-lbs.)
2855N models......................	1000 N·m
	(740 ft.-lbs.)
All other 2755 and 2955 models	1000 N·m
	(740 ft.-lbs.)

Torque can be measured using a pipe wrench and suitable handle extending pipe by dividing your total body weight into the desired torque and the result will be the distance from shaft center line to apply your weight. An example would be 740 ft.-lbs.(desired torque)/180 lbs. (your body weight) = 4.11 ft. (distance from shaft center line). To convert the distance to inches, multiply $4.9 \times 12 = 49.32$ inches.

   d. Mechanical failure in front-wheel drive.

2. Front-wheel drive not disengaging. Check by shifting transmission to NEUTRAL, then jacking up left front wheel. With key switch ON, dash "MFWD" switch OFF, engine operating at 1000 rpm and oil temperature warmed to at least 40° C (100° F), it should be possible to rotate the left front wheel by hand. Failure to disengage could be caused by:
   a. Electrical fault (switch, fuse or wiring). Check for magnetic field at solenoid valve using a screwdriver or similar object near the valve with key switch ON and dash "MFWD" switch OFF.
   b. Defective solenoid/hydraulic valve (21E or 21L—Fig. 25). The valve may stick if the valve attaching screws (12 or 27) are improperly or unevenly tightened. Refer to paragraphs 17 and 18 for tests.
   c. Hydraulic pressure too low or leaking piston seal rings. Pressure can be checked as outlined in paragraph 17.

3. Excessive front tire wear. Could be caused by:
   a. Toe-in incorrect.
   b. Front-rear tire combination not as specified.
   c. Incorrect tire inflation.

TIE ROD AND TOE-IN

All Models with MFWD

6. Tie rod ends may be one of several different types, but none are adjustable for wear and faulty units must be renewed.

To check toe-in, first turn steering wheel so that front wheels are in straight ahead position. Measure distance at front and rear of front wheels from rim flange to rim flange at hub height. Toe-in should be 2-5 mm ($^5/_{64}$-$^{13}/_{64}$ inch) for 2955 models, 0-3 mm (0-$^1/_8$ inch) for all other models with MFWD.

To adjust toe-in of 2855N models, remove cotter pin and slotted nut from tie rod end (C—Fig. 7), loosen clamp (B), then detach tie rod end from steering arm. Hold piston rod (A) and turn tie rod end as required to set toe-in at 0-3 mm (0-$^1/_8$ inch). Reattach tie rod end to steering arm and recheck toe-in after making changes. When adjustment is correct, tighten slotted nut to 95 N·m (70 ft.-lbs.) and install cotter pin. Adjust tie rod ends at each end of piston rod evenly. Tighten

clamp bolts (B) to a torque of 50 N·m (37 ft.-lbs.) when adjustment is correct.

To adjust toe-in of all other models with MFWD, loosen clamp bolt (F), then turn tie rod (E) in or out of tie rod end as required. Adjust both sides evenly. When adjustment is correct, tighten clamp bolt (F) of late 2750 models (APL-735) and all 2755 models to 30 N·m (22 ft.-lbs.) torque. Correct torque for clamp bolt on early 2750 models (APL-345) and all 2955 models is 55 N·m (40 ft.-lbs.).

DRIVE SHAFT

All Models with MFWD

7. To remove the drive shaft, first unbolt front and rear anti-wrapping guards and slide them toward center of drive shaft. Unbolt and separate yoke flange (1—Fig. 8) from drive hub on front differential pinion shaft. Pull drive shaft with antiwrapping guards (Fig. 9) forward off the clutch shaft and remove from tractor.

Check "U" joints for excessive wear and renew as necessary. "U" joint cross (3—Fig. 8), seals, bearing needles and bearing sleeves are available as an assembly with new snap rings (2). Removal and installation of "U" joints is conventional.

Reinstall drive shaft assembly by reversing removal procedure. Tighten yoke flange bolts to a torque of 75 N·m (55 ft.-lbs.).

FRONT DRIVE AXLE

All Models with MFWD

8. REMOVE AND REINSTALL. To remove the front drive axle and support assembly, remove drive shaft as outlined in paragraph 7. Disconnect power

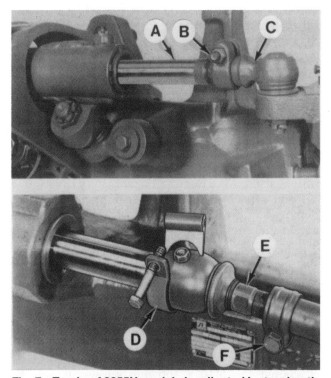

Fig. 7—Toe-in of 2855N models is adjusted by turning tie rod end (C) on piston rod (A). Toe-in of other models with MFWD is adjusted by turning tie rod (E) with 22 mm wrench. Be sure stop clamp (D) is correctly installed.

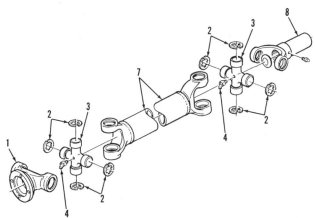

Fig. 8—Exploded view of front drive shaft and "U" joints used on models with front-wheel drive.

1. Yoke flange	4. Grease fitting
2. Snap rings	7. Drive shaft
3. "U" joint cross	8. Slip yoke

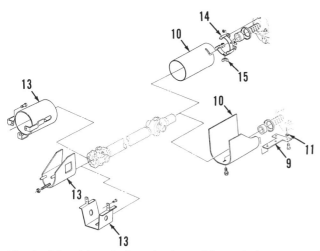

Fig. 9—The drive shaft to the front drive axle is protected by the anti-wrapping guard. Different styles have been used.

9. Bracket	13. Front guard
10. Rear guard	14. Bracket
11. Washer	15. Wing nuts

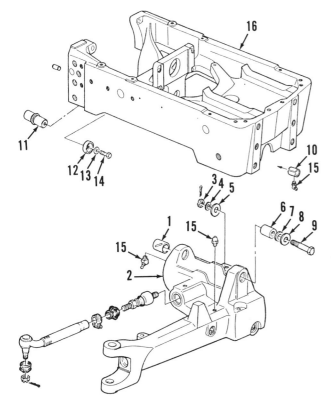

Fig. 10—Front drive axle pivots on pins (6 and 11) that move in bushings (1 and 10).

1. Rear bushing	9. Pivot bolt
2. Front drive axle	10. Front bushing
3. Slotted nut	11. Rear pivot tube
4. Washer	12. Bushing
5. Washer	13. Washer
6. Pivot tube	14. Cap screw
7. Shim	15. Grease fittings
8. Washer	16. Front axle support

steering lines at steering cylinder. Cap or plug openings immediately to prevent dirt from entering system. Support tractor behind axle carrier and remove front wheels. Support front axle assembly with a suitable jack or special safety stand to prevent tipping while permitting the axle to be moved safely. Refer to Fig. 10 and remove cotter pin, slotted nut (3) and washers (4 and 5). Remove special pivot bolt (9), washer (8) and shim (7). Move front axle rearward until clear of pivot tube (6) and rear pivot pin (11). Lower axle assembly and carefully roll away from under tractor.

Inspect bushing (10), front pivot tube (6), rear pivot pin (11) and bushings (1 and 10) for excessive wear and renew if necessary. Be sure to align passage in bushing with appropriate grease fitting if new bushing is installed. Crossed grooves of rear bushing (1) should be down and crossed grooves or slot in front bushing (10) should be up.

Reinstall front drive axle by reversing the removal procedure. Tighten slotted nut (3) to 300 N•m (220 ft.-lbs.) torque. End clearance of axle center member should be 0-0.5 mm (0-0.020 inch) and is adjusted by adding or removing shims (7). If cotter pin cannot be installed through slotted nut (3) after tightening to specified torque, tighten enough to align hole with next slot, then install cotter pin.

WHEEL HUB WITH PLANETARY

Early 2750, All 2855N and All 2955 Models So Equipped

9. OVERHAUL. Either front wheel hub and planetary can be serviced without removing the steering knuckle housing. Support front axle housing and remove front wheel. Remove drain plug (42—Fig. 11) and drain oil from hub assembly. Remove the two Allen screws (41) and lift off planetary carrier (40). Remove thrust washers (39), then remove snap rings (29). Using a suitable puller, remove planet gears (32) with bearings (31). Remove angular snap rings (30 or 33) and separate bearings (31) from planet gears. Remove snap ring (36) and thrust washer (34) from pinion shaft and sun gear (35).

Before removing the ring gear (28), mark relative position of ring gear (28) and steering knuckle housing (17) to facilitate installation in same location. Unscrew the eight self-locking cap screws (38). Install three M10 × 100 mm cap screws in holes around sun gear and using a disc plate (OTC 27545 or equivalent) against cap screw heads, attach a jawed puller to ring gear (28) with center pressing against disc plate that is against the M10 × 100 mm screws. It may be necessary to heat area of ring gear around bushings (37) to 300° C (570° F) and remove ring gear. Bushings (37) are installed with Loctite and heat must be used to release the Loctite. Use a slide hammer puller and remove bushings (37).

To remove the hub (25) and bearing cone (26) from steering knuckle (17), reinstall the three M10 × 100 mm cap screws around sun gear and use a disc plate (OTC 27545 or equivalent) against heads of screws. Attach a large jawed puller across hub with center pressing against disc plate that is against heads of screws and pull hub (25) from steering knuckle housing (17). Remove "O" ring (27) and cups for bearings (23 and 26) from hub, then remove inner bearing cone (23), oil seal (22) and scraper ring (21) from knuckle housing (17). If necessary, remove wear ring (20) from knuckle housing and bearing cups from hub.

Clean and inspect all parts for excessive wear or other damage and renew as necessary.

When reassembling, drive cups for bearings (23 and 26), oil seal (22) and scraper ring (21) in hub (25). If removed, install new wear ring (20) and cone for inner bearing (23) on knuckle housing (17). Install hub (25) and cone for outer bearing (26) on knuckle housing. It may be necessary to heat bearing cones to 120° C (250° F) before installing. Coat bushings (37) with Loctite 270 and drive into knuckle housing. If bushings (37) are spring-pin-type, split should be toward direction of rotation or away from direction of rotation. **Split of pin should never be toward center of axle or away from center of axle.** Use M10 × 100 mm cap screws and nuts to force ring gear on bushings. Secure ring gear with new self-locking cap screws (38), tightened to 90 N·m (65 ft.-lbs.) torque. Install thrust washer (34) and snap ring (36) on sun gear (35). Install new "O" ring (27) in groove

on hub (25). Install bearings (31) in planetary gears (32) and secure with angular snap rings (30 and 33). Install planetary gears on shafts and secure with snap rings (29).

Before installing planetary carrier, determine correct thickness thrust washer (39) to be used. Measure the distance outer end of sun gear (35) is from flange of hub (25). See Fig. 12. Next measure distance from flange of planetary carrier (40—Fig. 11) to thrust washer seat in carrier. From the difference of the two measurements, deduct 0.3-0.6 mm (0.012-0.023 inch) for the specified free play. The result is the correct thickness thrust washer to be installed. Use grease to stick thrust washer in place and install planetary carrier assembly. Install the two Allen screws (41). Fill hub and planetary to the level plug opening

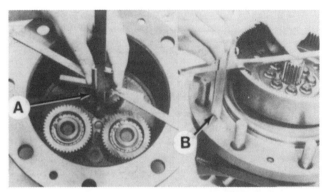

Fig. 12—Measure the distance "B" between end of pinion shaft (35—Fig. 11) and face of hub (25), then measure distance "A" between flange of planet carrier (40) and seated shims (39). Difference is end play. Change end play by varying thickness of shims (39).

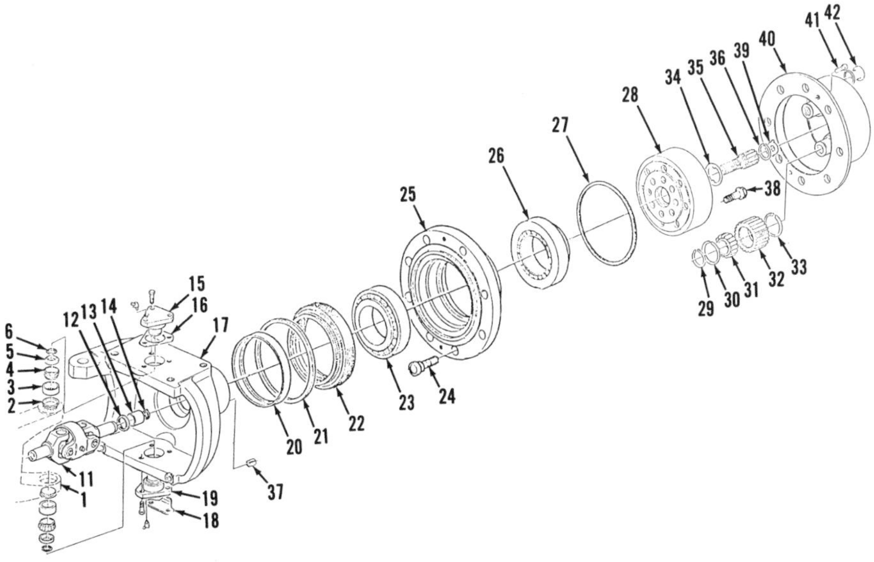

Fig. 11—Exploded view of APL-350 front wheel hub and planetary used on MFWD equipped 2955 models. The planetary unit on APL-345 axle, used on early 2750 models, and the APL-325N axle, used on 2855N models, are similar.

1. Axle
2. Cup
3. Bearing cup
4. Bearing cone
5. Seal cap
6. "O" ring
11. Double "U" joint
12. Oil seal
13. Bushing
14. Snap ring
15. King pin
16. Shim
17. Steering knuckle housing
18. Scraper plate

19. King pin
20. Wear ring
21. Scraper ring
22. Oil seal
23. Taper roller bearing
24. Lug bolt
25. Hub
26. Taper roller bearing

27. "O" ring
28. Ring gear
29. Snap ring
30. Angular snap ring
31. Roller bearing
32. Planetary gear
33. Angular snap ring
34. Thrust washer

35. Pinion shaft & sun gear
36. Snap ring
37. Bushing
38. Cap screw
39. Shim washers
40. Planet carrier
41. Allen head screws
42. Drain plug

(drain plug positioned so that "OLSTAND" mark is in horizontal position) with SAE 85W-140 GL-5 gear oil. Install front wheel and tighten nuts to a torque of 300 N•m (220 ft.-lbs.).

Late 2750 and All 2755 Models with MFWD

10. OVERHAUL. Either front wheel hub and planetary can be serviced without removing the steering knuckle housing. Support front axle housing and remove front wheel. Attach a suitable hoist to support hub assembly. Remove drain plug (42—Fig. 13) and drain oil from hub assembly. Remove the two Allen screws (41) and lift off planetary carrier (40). Remove snap rings (29), then use a suitable puller to remove planet gears (32) with bearings (31). Remove angular snap rings (30 or 33) and separate bearings (31) from planet gears.

Before removing ring gear (28), mark relative position of ring gear (28) and steering knuckle housing (17) to facilitate installation in same location. Remove cap screws (46) and lock tab (45). Use special wrench KJD 10147 or equivalent to remove nut (44), then remove ring gear (28). Use a jawed puller to remove hub (25) and bearings (23 and 26). Puller can push against center of sun gear shaft (35).

Clean and inspect all parts for excessive wear or other damage and renew as necessary.

When reassembling, drive bearing cups in hub, then install inner bearing cone, oil seal (22) and scraper ring (21) in hub (25). Pack cavities of oil seal and lubricate lips of seal with grease. If removed, install new wear ring (20) on knuckle housing. It may be necessary to heat bearing cones to 120° C (250° F) before installing. Slide ring gear (28) onto splines of knuckle housing and install nut (44) with beveled side out. Tighten nut (44) with special wrench KJD 10147 or equivalent to 450 N•m (330 ft.-lbs.) torque while turning the wheel hub. Rolling torque of hub is measured by attaching string to studs in hub and measuring with a spring scale. Correct rolling torque should be 50-62.5 N (11.2-14 lbs.) with oil seal (22) NOT installed. Be sure that bearings are seated when measuring, then install lock plate (45) and cap screws (46). It may be necessary to tighten nut (44) slightly to permit installation of cap screws (46). Tighten screws (46) to 40 N•m (30 ft.-lbs.) torque.

Install new "O" ring (27) in groove on hub (25). Install bearings (31) in planetary gears (32) and secure with angular snap rings (30 and 33). The wider side of snap rings (30 and 33) should be toward bearing rollers. Install planetary gears on shafts with greater radius of bearing inner race toward planet carrier, then secure with snap rings (29).

Before installing planetary carrier, determine correct thickness thrust washer (39) to be used. Measure the distance outer end of sun gear (35) is from flange of hub (25). See Fig. 12. Next measure distance from flange of planetary carrier (40—Fig. 13) to thrust washer seat in carrier. From the difference of the two measurements, deduct 0.3-0.6 mm (0.012-0.023 inch) for the specified free play. The result is the correct thickness thrust washer (or washers) to be installed. Position thrust washer of correct thickness in planetary carrier after coating back of thrust washer with Loctite 270. The carrier can also be staked with center punch to help retain the thrust washers. Install the planet carrier assembly and the two Allen screws (41). Fill hub and planetary to the level plug opening (drain plug positioned so that "OLSTAND" mark is in horizontal position) with SAE 85W-140 GL-5 gear oil. Install front wheel and tighten nuts to a torque of 300 N•m (220 ft.-lbs.).

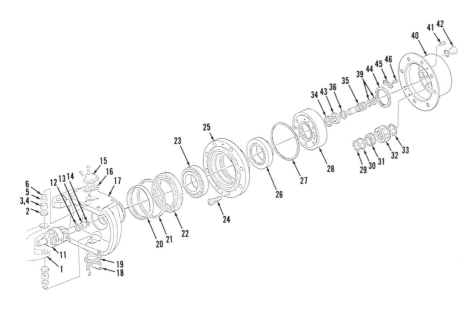

Fig. 13—Exploded view of APL-735 front wheel hub and planetary used on late 2750 and all 2755 models with MFWD. Refer to Fig. 11 for legend except the following:

34. Thrust washer
35. Pinion shaft & sun gear
36. Snap ring
39. Shims
40. Planet carrier
41. Allen head screws
42. Drain plug
43. Thrust washer
44. Retaining nut
45. Lock plate
46. Cap screws

STEERING KNUCKLE HOUSING

All Models with MFWD

11. R&R AND OVERHAUL. To remove either steering knuckle housing, first remove wheel hub and planetary as outlined in paragraph 9 or 10. Disconnect tie rod from steering knuckle arm. Unbolt and remove upper and lower king pins (15 and 19—Fig. 14 or Fig. 19) and scraper plate (18), if so equipped. Measure and note thickness and number of shims (16) at upper and lower king pin for aid in reassembly. Carefully remove steering knuckle housing (17). Axle shaft and double "U" joint assembly may be removed with knuckle housing. Lower bearing cone (4), seal cap (5) and "O" ring (6) will fall from axle housing fork (1). Remove upper bearing cone, seal cap and "O" ring. Bearing cups (3) and grease cups (2) can be removed if necessary.

If desired, axle shaft and "U" joint assembly can be withdrawn for inspection or repair. If renewal is required, oil seal (12) and bushing (13) can be removed from knuckle housing (17). Oil seal (8) and bushing (7) can be removed from axle housing (1). Bushings (7 and 13) should be pressed into position with external groove toward top and internal arrow-shaped grooves pointing toward inside of oil-filled housing (away from seal). Using snap ring pliers, open snap ring (10) and pull axle shaft from "U" joint. Spread snap ring (14) and remove sun gear (35—Fig. 11 or Fig. 13) from "U" joint. "U" joint kits are available and installation procedure is conventional. Be careful not to damage seals (8 and 12—Fig. 14 or Fig. 19) when installing axle and knuckle housing.

Reassemble by reversing the disassembly procedure. Tighten king pin retaining cap screws to a torque of 120 N·m (85 ft.-lbs.). Using a torque wrench on upper king pin cap screw, check rolling drag torque of king pin bearings. Add or remove shims (16), equally at top and bottom, as required to obtain rolling drag torque of 7-8 N·m (5-6 ft.-lbs.) for 2855N models (APL-325N axle), 12-14 N·m (9-10 ft.-lbs.) for early 2750 and all 2955 models (APL-345 and APL-350 axles), 9-11 N·m (6.5-8 ft.-lbs.) for later 2750 models and all 2755 models (APL-735 axles). Shims are available in varying thicknesses.

Refer to paragraph 9 or 10 when reassembling hub and planetary.

DIFFERENTIAL UNIT

Early 2750 and All 2955 Models with MFWD

NOTE: Although the differential can be removed with right axle housing (47—Fig. 14) attached to tractor, most mechanics prefer removing complete

1. Axle left housing
2. Cup
3. Bearing cup
4. Bearing cone
5. Seal cap
6. "O" ring
7. Bushing
8. Oil seal
9. Axle shaft
10. Snap ring
11. Double "U" joint
12. Oil seal
13. Bushing
14. Snap ring
15. King pin
16. Shim
17. Steering knuckle housing
18. Scraper plate
19. King pin
47. Axle right & center housing
48. "O" ring
49. Shims
50. Taper roller bearing
51. Level plug
52. Drain plug
53. Tie rod end
54. Boot
55. Clamp
56. Tie rod
57. Nut

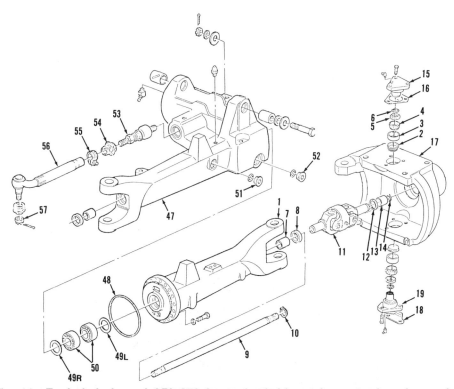

Fig. 14—Exploded view of APL-350 front-wheel drive axle center housing and steering knuckle used on MFWD equipped 2955 models. Parts are similar for APL-345 axle used on early 2750 models and for APL-735 axle used on later 2750 and all 2755 models. Refer to Fig. 19 for APL-325N axle used on 2855N models.

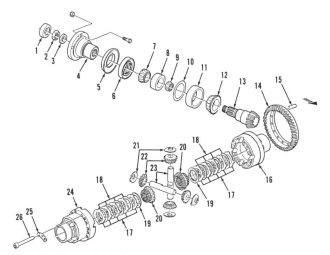

Fig. 15—Exploded view of front drive axle drive pinion and differential unit used on early 2750 models and all 2955 models.

1. Lock plate	14. Ring gear
2. Nut	15. Dowel pin
3. Washer	16. Differential housing
4. Drive hub	17. External lugged plates
5. Dust shield	18. Internal splined discs
6. Oil seal	19. Compensating discs
7. Bearing cone	20. Bevel side gears
8. Bearing cup	21. Thrust washers
9. Crush spacer	22. Bevel pinions
10. Shims	23. Pinion shaft
11. Bearing cup	24. Differential case
12. Bearing cone	25. Plate
13. Drive pinion	26. Special screws

Fig. 16—Refer to text and these illustrations when checking and adjusting carrier bearing preload.

front drive axle assembly from tractor and placing it on stands or a large bench.

12. R&R AND OVERHAUL. To remove the differential, first remove axle assembly as outlined in paragraph 8. Although not required, some mechanics prefer to remove both hub and planetary units as in paragraph 9 and steering knuckle housings and axle shafts as in paragraph 11. Remove drain plug (52—Fig. 14) and drain oil from axle housing. Unbolt and remove left axle housing (1). Lift out ring gear and differential unit. Remove cups for bearings (50) and shims (49L and 49R) from each axle housing. Shims (49L and 49R) are used to adjust backlash of ring gear and pinion, and preload of carrier bearings. Measure and record thickness of shims (49L and 49R) for each side when disassembling to facilitate reassembly and adjustment. Remove "O" ring (48) from housing (1).

To remove the drive pinion (13—Fig. 15), unlock lock washer (1) and remove nut (2), washer (3), drive hub (4) and shield (5), if so equipped. Drive the pinion shaft (13) inward using a soft hammer and remove pinion and related parts from axle housing. Remove seal (6) and bearing cone (7). Use a slide hammer puller to remove bearing cup (8). Drive bearing cup (11) out of bore and remove from inside the housing. Measure and note thickness and number of shims (10). Remove crush spacer (9) and use suitable puller to remove bearing cone (12) from drive pinion shaft.

Remove the twelve socket-head screws (26) and separate case half (24) from case half (16). Remove pinion shafts (23), bevel pinions (22) and thrust washers (21), then remove side gears (20), compensating discs (19) and external and internal splined discs (17 and 18) from differential case halves (16 and 24). Using a suitable puller, remove ring gear (14) from case half (16). Working through interior holes, use a hammer and punch to drive taper roller bearing cones from case halves.

Ring gear (14) and drive pinion (13) are available only as a matched set. If all original parts are reinstalled, original shims (10—Fig. 15) and shims (49L and 49R—Fig. 14) can be reused. However, if bearings, differential case, gears or housings are renewed, the correct thickness for new shim packs must be determined. Clean and inspect all parts and renew any showing excessive wear or other damage.

Assemble differential unit as follows: Drive spring pin alignment dowels (15—Fig. 15) into case half (16) with split toward direction of rotation, either front or rear. **DO NOT have split toward center or toward outer diameter of ring gear.** Heat ring gear (14) to 120° C (250° F) and install over spring pins on case half (16). Heat cones for bearings (50—Fig. 14) to 120° C (250° F) and install on case halves (16 and 24—Fig. 15). Install compensating discs (19) and internal and external splined discs (17 and 18) on side gears (20), then install side gear assemblies in case halves (16 and 24). Assemble bevel pinions (22) and thrust washers (21) on pinion shafts (23), then place assembly in case half (16) over side gear. Make sure that bevel pinion shafts (23) are tight against case half (16), then use a dial indicator to measure backlash between bevel pinions (22) and side gear (20). Specified backlash is 0.2-0.3 mm (0.007-0.011 inch). Add or remove compensating discs (19) or install compensating discs of different thicknesses as required to obtain correct backlash. Transfer bevel pinions, shafts and thrust washers to opposite case half (24) and check backlash between bevel pinions and other side gear. Change compensating disc thicknesses as required to set backlash within the correct range of 0.2-0.3 mm (0.008-0.0122 inch). Coat all parts of the disc packs, thrust washers, pinions and shafts with gear oil before final assembly. Assemble case halves so that stamped numbers are opposite each other. Install the twelve socket-head cap screws and tighten evenly, using a crossing pattern, to 160 N•m (120 ft.-lbs.) torque. Heat bearing cones and inner races for bearings (50—Fig. 14) to 120° C (250° F) and push into position. Be sure races are seated against shoulders.

The correct thickness of shim pack (10—Fig. 15) is determined as follows: The measurement from the axle centerline to the bottom of bore in housing for bearing cup (11) is stamped by the manufacturer in millimeters on right axle housing just above the axle model plate as shown in Fig. 18. Place bearing cone (12—Fig. 15) in cup (11) and measure total height of bearing. Add this measurement to dimension stamped in millimeters on face of drive pinion (13). Subtract this sum from dimension stamped in axle housing. The difference is the required thickness of shim pack (10). Place shim pack in bearing bore and install bearing cup (11). Install outer bearing cup (8). Heat bearing cone (12) and install on pinion shaft (13). Slide crush spacer (9) on pinion shaft, then install pinion shaft in place in axle housing. Heat bearing cone (7) and install on pinion shaft. Coat lips of seal (6) with grease and install seal in housing bore. Install drive hub (4), washer (3) and nut (2). Tighten nut (2) to 260 N•m (190 ft.-lbs.) torque and measure rolling torque of pinion shaft in bearings. The drive pinion shaft rolling drag torque should be 2-3 N•m (18-26 in.-lbs.) when nut (2) is correctly tightened. Secure nut with lock plate (1).

Adjust ring gear backlash and rolling drag torque as follows: Install one shim (49R—Fig. 14) of 1 mm (0.040 inch) thickness in right axle housing (47). Install bearing cup and make sure cup is seated against shim. Place differential assembly in right axle housing. Check backlash between ring gear and drive pinion using a dial indicator through oil drain hole. Backlash as tested (without preload on bearing) should be 0.2-0.3 mm (0.008-0.012 inch). Use differ-

ent thickness shim behind bearing cup to correct backlash, if necessary.

Place bearing cup on left differential bearing cone and measure distance from right axle housing flange to bearing cup surface as shown in top illustration Fig. 16. Using spacer blocks and a straightedge, measure distance from left axle housing flange to bearing cup seat as shown in lower illustration in Fig. 16. Subtract the second measurement from the first measurement. This difference plus 0.2 mm (0.008 inch) will be the correct thickness shim to be used to obtain a rolling drag torque of 1-3 N•m (9-24 in.-lbs.).

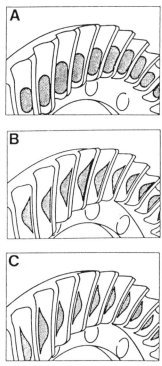

Fig. 17—Views of typical tooth contact patterns on ring gear.

   A. Ideal tooth contact pattern
   B. Drive pinion must be moved toward ring gear
   C. Drive pinion must be moved away from ring gear

Fig. 18—Arrow indicates the location of dimension stamped on axle housing that is used when setting pinion position.

Install this shim (49L—Fig. 14) and bearing cup in left axle housing.

Tooth contact pattern can be checked on all models as follows: Lift differential assembly from right axle housing and coat several ring gear teeth with marking blue. Reinstall differential assembly and using four cap screws, attach axle housings together. Turn drive pinion shaft in both directions several times. Remove left axle housing and lift out differential assembly. Inspect tooth contact pattern and refer to Fig. 17. If ideal contact pattern is not made, recheck all adjustments.

Complete reassembly as follows: Position differential assembly in right housing (47—Fig. 14), install new "O" ring (48) in groove of left axle housing (1), then install left axle housing, tightening retaining cap screws to a torque of 190 N·m (140 ft.-lbs.). Reinstall axle shafts and steering knuckle housings as outlined in paragraph 11 and wheel hub and planetary units as in paragraph 9. Reinstall front drive axle assembly as in paragraph 8. Fill wheel hubs and front axle housing to level plug openings with SAE 85W-140 GL-5 gear oil. Capacity for each wheel hub is approximately 0.75 L (0.8 quart) for early 2750 models; 1.0 L (1.04 quarts) for 2955 models. Capacity for axle housing is approximately 5.3 L (5.6 quarts) for early 2750 models; 6.5 L (6.9 quarts) for 2955 models.

2855N Models with MFWD

NOTE: Although the differential can be removed with right axle housing (47—Fig. 19) attached to tractor, most mechanics prefer removing complete front drive axle assembly from tractor and placing it on stands or a large bench.

13. R&R AND OVERHAUL. To remove the differential, first remove axle assembly as outlined in paragraph 8. Although not required, some mechanics prefer to remove both hub and planetary units as in paragraph 9 and steering knuckle housings and axle

shafts as in paragraph 11. Remove drain plug (52—Fig. 19) and drain oil from axle housing. Unbolt and remove left axle housing (1). Lift out ring gear and differential unit. Remove cups for bearings (50) and shims (49L and 49R) from each axle housing. Shims (49L and 49R) are used to adjust backlash of ring gear and pinion and preload of carrier bearings. Measure and record thickness of shims (49L and 49R) for each side when disassembling to facilitate reassembly and adjustment. Remove "O" ring (48) from housing (1).

To remove the drive pinion (13—Fig. 20 or Fig. 21), unlock lock plate (1—Fig. 20), if so equipped, and remove nut (2—Fig. 20 or Fig. 21), washer (3) and drive hub (4). Drive the pinion shaft (13) inward using a soft hammer and remove pinion and related parts from axle housing. Remove seal (6) and bearing cone (7). Use a slide hammer puller to remove bearing cup (8). Drive bearing cup (11) out of bore and remove from inside the housing. Measure and note thickness and number of shims (10). Remove crush spacer (9) and use suitable puller to remove bearing cone (12) from drive pinion shaft.

Ring gear (14—Fig. 20 or Fig. 21) and drive pinion (13) are available only as a matched set. If all original parts are reinstalled, original shims (10) and shims (49L and 49R—Fig. 19) can be reused. However, if bearings, differential case, gears or housings are renewed, new shim packs must be determined. Clean and inspect all parts and renew any showing excessive wear or other damage.

Refer to paragraph 14 for disassembly, inspection and repair of self-locking differential used on some models. The following includes service procedures common to both locking and non-locking types. Application will be noted.

On non-locking differential, use a suitable bearing puller to remove carrier bearing cone from ring gear (14—Fig. 20). Pry lock plate (29) from ring gear, then remove ten screws (28) attaching ring gear to differential housing (24). Screw three M10 × 80 mm screws into differential case in place of removed screws (28),

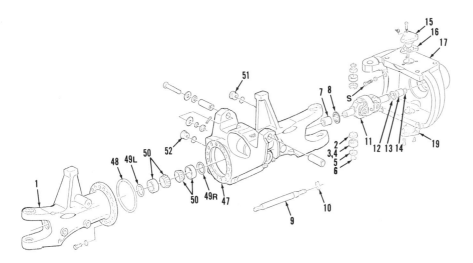

Fig. 19—Exploded view of APL-325N front-wheel drive axle center housing and steering knuckle used on 2855N models. Refer to Fig. 14 for legend.

position a disc (such as OTC 27549 or equivalent) on heads of the long screws and use a suitable bearing puller to push against the disc and heads of screws to separate ring gear (14) from case (24). Remove the puller, disc and the temporarily installed long screws, then remove ring gear. Use a 5 mm ($^{13}/_{64}$ inch) diameter punch through one of the threaded holes in case (24) to drive pin (27) out of pinion shaft (23). **DO NOT attempt to drive pin into the threads.** After pinion shaft (23) is withdrawn, side gears (20), pinions (22) and thrust washers (30) can be removed.

Assembly of non-locking differential is reverse of disassembly procedure. Roll pins (15) should be driven into position with split toward direction of rotation, either front or rear. **DO NOT have split toward center or toward outer diameter of ring gear.** Tighten screws (28) evenly using a crossing pattern to a final torque of 70 N•m (50 ft.-lbs.); then, install lock plate (29). Heat bearing cones and inner races for bearings (50—Fig. 19) to 120° C (250° F) and push into position. Be sure races are seated against shoulders.

On all models of differential, the correct thickness of shim pack (10—Fig. 20 or Fig. 21) is determined as follows: The measurement from the axle centerline to the bottom of bore in housing for bearing cup (11) is stamped in millimeters by the manufacturer on right axle housing just above the axle model plate. Place bearing cone (12) in cup (11) and measure total height of bearing. Add this measurement to dimension stamped in millimeters on face of drive pinion (13). Subtract this sum from dimension stamped in axle housing. The difference is the required thickness of

shim pack (10). Place shim pack in bearing bore and install bearing cup (11). Install outer bearing cup (8). Heat bearing cone (12) and install on pinion shaft (13). Slide crush spacer (9) on pinion shaft, then install pinion shaft in place in axle housing. Heat

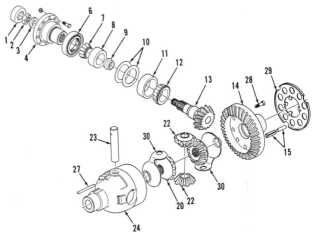

Fig. 20—Exploded view of front drive axle and differential unit used on some 2855N models with non-locking differential. Refer to Fig. 21 for self-locking differential.

1. Lock plate	13. Drive pinion & shaft
2. Nut	14. Ring gear
3. Washer	15. Dowel pins
4. Drive hub	20. Side gears
6. Oil seal	22. Bevel pinions
7. Bearing cone	23. Pinion shaft
8. Bearing cup	24. Differential case
9. Crush spacer	27. Pin
10. Shims	28. Special screws
11. Bearing cup	29. Lock plate
12. Bearing cone	30. Thrust washers

2. Nut
3. Washer
4. Drive hub
6. Oil seal
7. Bearing cone
8. Bearing cup
9. Crush spacer
10. Shims
11. Bearing cup
12. Bearing cone
13. Drive pinion & shaft
14. Ring gear
15. Dowel pins
17. External lugged plates
18. Internal splined discs
19. Thrust washers
20. Side gears
21. Thrust washers
22. Bevel pinions
23. Pinion shaft
24. Differential case
27. Pin
28. Special screws
29. Lock plate

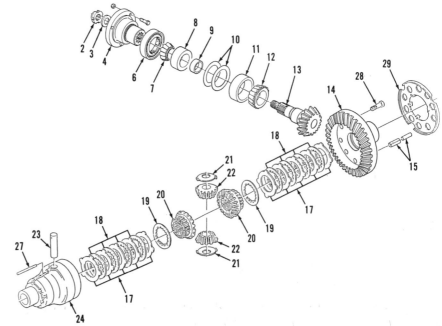

Fig. 21—Exploded view of self-locking front drive differential used on some models.

bearing cone (7) and install on pinion shaft. Coat lips of seal (6) with grease and install seal in housing bore. Install drive hub (4), washer (3) and nut (2). Tighten nut (2) to 200 N·m (150 ft.-lbs.) torque and measure rolling torque of pinion shaft in bearings. The drive pinion shaft rolling drag torque should be 2-3 N·m (18-26 in.-lbs.) when nut (2) is correctly tightened. Secure nut with lock plate (1—Fig. 20), if so equipped.

Adjust ring gear backlash and rolling drag torque of all models as follows: Install one shim (49R—Fig. 19) of 1 mm (0.040 inch) thickness in right axle housing (47). Install bearing cup and make sure cup is seated against shim. Place differential assembly in right axle housing. Using a dial indicator through oil drain hole, check backlash between ring gear and drive pinion. Backlash should be, under this condition (without preload on bearing), 0.2-0.3 mm (0.008-0.012 inch). If necessary, use different thickness shim behind bearing cup to correct backlash.

Place bearing cup on left differential bearing cone and measure distance from right axle housing flange to bearing cup surface as shown in top illustration Fig. 16. Using spacer blocks and a straightedge, measure distance from left axle housing flange to bearing cup seat as shown in lower illustration in Fig. 16. Subtract the second measurement from the first measurement. This difference plus 0.2 mm (0.008 inch) will be the correct thickness shim to be used to obtain a rolling drag torque of 1-3 N·m (9-24 in.-lbs.). Install this shim (49L—Fig. 19) and bearing cup in left axle housing (1).

Tooth contact pattern can be checked on all models as follows: Lift differential assembly from right axle housing and coat several ring gear teeth with marking blue. Reinstall differential assembly and using four cap screws, attach axle housings together. Turn drive pinion shaft in both directions several times. Remove left axle housing and lift out differential assembly. Inspect tooth contact pattern and refer to Fig. 17. If ideal contact pattern is not made, recheck all adjustments.

Complete reassembly of all types of differentials as follows: Position differential assembly in right hous-

Fig. 22—Clearance of inner clutch pack can be measured using a feeler gage through hole as shown. Refer to text for procedure.

ing (47—Fig. 19), install new "O" ring (48) in groove of left axle housing (1), then install left axle housing, tightening retaining cap screws to a torque of 190 N·m (140 ft.-lbs.). Reinstall axle shafts and steering knuckle housings as outlined in paragraph 11 and wheel hub and planetary units as in paragraph 9. Reinstall front drive axle assembly as in paragraph 8. Fill wheel hubs and front axle housing to level plug openings with SAE 85W-140 GL-5 gear oil. Capacity for each wheel hub is 0.75 L (0.8 quart) and for axle housing is 5.3 L (5.6 quarts).

14. SELF-LOCKING DIFFERENTIAL. Use a suitable bearing puller to remove carrier bearing cone from ring gear (14—Fig. 21). Pry lock plate (29) from ring gear, then remove ten screws (28) attaching ring gear to differential housing (24). Screw three or four M10 × 80 mm screws into differential case in place of removed screws (28), position a disc (such as OTC 27549 or equivalent) on heads of the long screws and use a suitable bearing puller to push against the disc and heads of screws to separate ring gear (14) from case (24). Remove the puller, disc and the temporarily installed long screws, then remove ring gear. Lift out the left side gear (20) and clutch pack (17, 18 and 19) from the open side and keep these parts separate from similar parts for opposite side. Use a 5 mm (13/64 inch) diameter punch through one of the threaded holes in case (24) to drive pin (27) out of pinion shaft (23). **DO NOT attempt to drive pin into the threads.** After pinion shaft (23) is withdrawn, side gear (20), pinions (22) and thrust washers (21) can be removed. The right side gear (20) and clutch pack (17, 18 and 19) can then be removed. Although parts (17, 18 and 19) are alike for both sides, parts should be kept separate if they are to be reinstalled. The tapered roller bearing cone can be driven from differential case using a punch through interior holes in case.

Assemble differential as follows: If previously used parts are to be reinstalled, they should be located in their original locations or abnormal wear may result. Thickness of clutch packs (17, 18 and 19) should be exactly the same thickness on both sides of pinion gears (20). When measuring thickness of clutch packs and measuring depth of clutch pack in differential case, parts should be **dry**.

Assemble alternating five external lugged plates (17) and four internal splined discs (18), then install one thrust washer (19) and side gear (20) in differential case. Position thrust washers (19) and pinion gears (20) in case, then insert shaft through case, thrust washers and pinion gears. Use a feeler gage through hole in differential case as shown in Fig. 22 to measure clearance between first external lugged plate (17—Fig. 21) and surface of case. Correct clearance is 0.1-0.2 mm (0.004-0.008 inch). The external lugged plates (17) are available in varying thicknesses and the correct combination of plates should be installed to provide specified clearance. Be sure that measurement is with dry parts and that external plates (17) and internal splined discs (18) are alternated. When clearance is correct, lubricate parts and install pin (27) to hold pinion shaft. Assemble side gear (20) and similar clutch pack (17, 18 and 19) in case and measure depth from flange of case to the last external lugged plate (17). Depth measured is the clutch pack operating clearance and should be as near as possible to clearance of pack on other side. Roll pins (15) should be driven into position with split toward direction of rotation, either front or rear. **DO NOT have split toward center or toward outer diameter of ring gear.** Tighten screws (28) evenly using a crossing pattern to a final torque of 70 N·m (50 ft.-lbs.); then, install lock plate (29). Heat bearing cones and inner races for bearings (50—Fig. 19) to 120° C (250° F) and push into position. Be sure that races are seated against shoulders.

Refer to paragraph 13 for remainder of assembly, adjustment and installation procedures.

Late 2750 and All 2755 Models with MFWD

NOTE: Although the differential can be removed with right axle housing (47—Fig. 14) attached to tractor, most mechanics prefer removing complete front drive axle assembly from tractor and placing it on stands or a large bench.

15. R&R AND OVERHAUL. To remove the differential, first remove axle assembly as outlined in paragraph 8. Although not required, some mechanics prefer to remove both hub and planetary units as in paragraph 10 and steering knuckle housings and axle shafts as in paragraph 11. Remove drain plug (52—Fig. 14) and drain oil from axle housing. Unbolt and remove left axle housing (1). Lift out ring gear and differential unit. Remove cups for bearings (50) and shims (49L and 49R) from each axle housing. Shims (49L and 49R) are used to adjust backlash of ring gear and pinion and preload of carrier bearings. Measure and record thickness of shims (49L and 49R) for each side when disassembling to facilitate reassembly and adjustment. Remove "O" ring (48) from housing (1).

To remove the drive pinion (13—Fig. 21) remove nut (2), washer (3) and drive hub (4). Drive the pinion shaft (13) inward using a soft hammer and remove pinion and related parts from axle housing. Remove seal (6) and bearing cone (7). Use a slide hammer puller to remove bearing cup (8). Drive bearing cup (11) out of bore and remove from inside the housing. Measure and note thickness and number of shims (10). Remove crush spacer (9) and use suitable puller to remove bearing cone (12) from drive pinion shaft.

Ring gear (14) and drive pinion (13) are available only as a matched set. If all original parts are rein-

stalled, original shims (10—Fig. 21) and shims (49L and 49R—Fig. 14) can be reused. However, if bearings, differential case, gears or housings are renewed, correct thicknesses for new shim packs must be determined. Clean and inspect all parts and renew any showing excessive wear or other damage.

Use a suitable bearing puller to remove carrier bearing cone from ring gear (14—Fig. 21). Pry lock plate (29) from ring gear, then remove ten screws (28) attaching ring gear to differential housing (24). Screw three or four M10 × 80 mm screws into differential case in place of removed screws (28), position a disc (such as OTC 27549 or equivalent) on heads of the long screws and use a suitable bearing puller to push against the disc and heads of screws to separate ring gear (14) from case (24). Remove the puller, disc and the temporarily installed long screws, then remove ring gear. Lift out the side gear (20) and clutch pack (17, 18 and 19) from the open side and keep these parts separate from similar parts for opposite side. Use a 5 mm ($^{13}/_{64}$ inch) diameter punch through one of the threaded holes in case (24) to drive pin (27) from pinion shaft (23). **DO NOT attempt to drive pin into the threads.** After pinion shaft (23) is withdrawn, side gears (20), pinions (22) and thrust washers (21) can be removed. Although parts (17, 18 and 19) are alike for both sides, parts should be kept separate if they are to be reinstalled. The tapered roller bearing cone can be driven from differential case using a punch through interior holes in case.

Assemble differential as follows: If previously used parts are to be reinstalled, they should be located in their original locations or abnormal wear may result. Thickness of clutch packs (17, 18 and 19) should be exactly the same thickness on both sides of pinion gears (20). When measuring thickness of clutch packs and measuring depth of clutch pack in differential case, parts should be **dry**.

Alternate five external lugged plates (17) and four internal splined discs (18), then install one thrust washer (19) and side gear (20) in differential case. Position thrust washers (21) and pinion gears (22) in case, then insert shaft through case, thrust washers and pinion gears. Use a feeler gage through hole in differential case as shown in Fig. 22 to measure clearance between first external lugged plate (17—Fig. 21) and surface of case. Correct clearance is 0.1-0.2 mm (0.004-0.008 inch). The external lugged plates (17) are available in varying thicknesses and the correct combination of plates should be installed to provide specified clearance. Be sure that measurement is with dry parts and that external plates (17) and internal splined discs (18) are alternated. When clearance is correct, lubricate parts and install pin (27) to hold pinion shaft. Assemble side gear (20) and similar clutch pack (17, 18 and 19) in case and measure depth from flange of case to the last external lugged plate (17). Depth measured is the clutch pack operating clearance and should be as near as possible to clearance of pack on other side. Roll pins (15) should be driven into position with split toward direction of rotation, either front or rear. **DO NOT have split toward center or toward outer diameter of ring gear.** Tighten screws (28) evenly using a crossing pattern to a final torque of 70 N·m (50 ft.-lbs.); then, install lock plate (29). Heat bearing cones and inner races for bearings (50—Fig. 14) to 120° C (250° F) and push into position. Be sure that races are seated against shoulders.

Install the drive bevel pinion and determine correct thickness of shim pack (10—Fig. 21) as follows: The measurement from the axle centerline to the bottom of bore in housing for bearing cup (12) is stamped by the manufacturer in millimeters on right axle housing just above the axle model plate. Place bearing cone (12) in cup (11) and measure total height of bearing. Add this measurement to dimension stamped in millimeters on face of drive pinion (13). Subtract this sum from dimension stamped in axle housing. The difference is the required thickness of shim pack (10). Place shim pack in bearing bore and install bearing cup (11). Install outer bearing cup (8). Heat bearing cone (12) and install on pinion shaft (13). Slide crush spacer (9) on pinion shaft, then install pinion shaft in place in axle housing. Heat and install bearing cone (7). Coat lips of seal (6) with grease and install seal in housing bore. Install drive hub (4), washer (3) and nut (2). Tighten nut (2) to 200 N·m (150 ft.-lbs.) torque and measure rolling torque of pinion shaft in bearings. The drive pinion shaft rolling drag torque should be 2-3 N·m (18-26 in.-lbs.) when nut (2) is correctly tightened.

Adjust ring gear backlash and rolling drag torque as follows: Install one shim (49R—Fig. 14) of 1 mm (0.040 inch) thickness in right axle housing (47). Install bearing cup and make sure cup is seated against shim. Place differential assembly in right axle housing. Using a dial indicator through oil drain hole, check backlash between ring gear and drive pinion. Backlash should be, under this condition (without preload on bearing), 0.2-0.3 mm (0.008-0.012 inch). If necessary, use different thickness shim behind bearing cup to correct backlash.

Place bearing cup on left differential bearing cone and measure distance from right axle housing flange to bearing cup surface as shown in top illustration Fig. 16. Using spacer blocks and a straightedge, measure distance from left axle housing flange to bearing cup seat as shown in lower illustration in Fig. 16. Subtract the second measurement from the first measurement. This difference plus 0.2 mm (0.008 inch) will be the correct thickness shim to be used to obtain a rolling drag torque of 1-3 N·m (9-24 in.-lbs.). Install this shim and bearing cup in left axle housing.

Tooth contact pattern can be checked on all models as follows: Lift differential assembly from right axle

housing and coat several ring gear teeth with marking blue. Reinstall differential assembly and using four cap screws, attach axle housings together. Turn drive pinion shaft in both directions several times. Remove left axle housing and lift out differential assembly. Inspect tooth contact pattern and refer to Fig. 17. If ideal contact pattern is not made, recheck all adjustments.

Complete reassembly as follows: Position differential assembly in right housing (47—Fig. 14), install new "O" ring (48) in groove of left axle housing (1), then install left axle housing, tightening retaining cap screws to a torque of 190 N·m (140 ft.-lbs.). Reinstall axle shafts and steering knuckle housings as outlined in paragraph 11 and wheel hub and planetary units as in paragraph 10. Reinstall front drive axle assembly as in paragraph 8. Fill wheel hubs and front axle housing to level plug openings with SAE 85W-140 GL-5 gear oil. Capacity for each wheel hub is 0.75 L (0.8 quart) and for axle housings is 4.6-5.0 L (4.9-5.3 quarts).

FRONT-WHEEL DRIVE CLUTCH

All Models with MFWD

16. R&R AND OVERHAUL. To remove the front-wheel drive disc clutch assembly, first remove drain plug and drain transmission oil. Disconnect wires from solenoid. Unbolt and remove oil pan and solenoid assembly from clutch housing. Remove the drive shaft as outlined in paragraph 7. Remove plastic sleeve (22—Fig. 23) and washer (23) from front end of shaft (24). Unbolt and remove bearing quill (30) with seal washer (29), oil seal (28), "O" ring (27) and shim (26). Install a cap screw and washer in front end of clutch shaft. Attach a slide hammer puller and pull clutch shaft forward until front bearing cup is free of clutch housing. Support disc clutch assembly with one hand and pull clutch shaft out the front.

NOTE: When clutch shaft is removed, cone for rear bearing (1) and spacer (2) are free to drop out of clutch housing.

Lower disc clutch assembly carefully from clutch housing and catch bearing cone (1) and spacer (2).

To disassemble the clutch, refer to Fig. 23, then unbolt and remove covers (3 and 20). Pull clutch hub (5) from clutch assembly. Tilt clutch assembly and remove thrust washer (16). Use a JDT-24A spring compressor or place assembly in a press and apply pressure to Belleville springs, then remove snap ring (6). Release pressure and remove Belleville springs (7), rear pressure ring (8), internal and external splined discs (9 and 10) and center pressure ring (11). **Different numbers of internal splined discs (9) and external splined discs (10) are used** depending upon tractor model, serial number range and specific configuration. Be sure to assemble using the correct quantity. Additionally, some early 2855N models may be equipped with a spacer (10A) and the quantity of internal and external discs are each reduced by 1 with spacer installed.

Remove snap ring (12), spacer ring (13) and front pressure ring (8). Compressed air can be directed through hole in center bore of clutch drum (18) to begin forcing piston (14) with piston ring (15) from bottom of clutch drum. Use screwdrivers to carefully guide piston ring (15) past grooves on clutch drums. Remove front bearing cone (25) and "O" rings (19) from clutch shaft. Remove rear bearing cup (1) from clutch housing, if necessary.

Clean and inspect all parts and renew any showing excessive wear or other damage. Renew all "O" rings and seal rings when reassembling. Press new bushing (4) in clutch hub (5), if necessary.

To reassemble clutch, insert seal ring (17) in groove of clutch drum (18). Install piston ring (15) on piston (14), then install piston in clutch drum. Install front pressure ring (8), release ring (13) and secure with snap ring (12). Rounded edge of snap ring (12) should be toward pressure ring (11). Install center pressure ring (11) with chamfered side facing snap ring. On early models, install one 2.5 mm (0.098 inch) thick external splined disc (10A) as a spacer. Starting with internal splined disc (9), alternately install internal splined and external splined discs (10).

The correct quantity of internal splined discs (9) and external splined discs (10) is carefully matched to the tractor equipment and model. Changes are not recommended. Refer to the following for original application:

All 2750 tractors are originally equipped with a spacer ring (13) that is 34.2 mm (1.35 inches) thick, nine internal splined plates (9) and eight externally splined plates (10).

Model 2755 (not high-clearance) tractors before serial number 655825 are originally equipped with a spacer ring (13) that is 34.2 mm (1.35 inches) thick, nine internal splined plates (9) and eight externally splined plates (10).

Model 2755 (not high-clearance) tractors after serial number 655824 are originally equipped with a spacer ring (13) that is 37 mm (1.46 inches) thick, ten internal splined plates (9) and nine externally splined plates (10).

Model 2755 high-clearance models before serial number 646531 are originally equipped with a spacer ring (13) that is 34.2 mm (1.35 inches) thick, nine internal splined plates (9) and eight externally splined plates (10).

Model 2755 high-clearance tractors after serial number 646530 are originally equipped with a spacer ring (13) that is 38.3 mm (1.5 inches) thick,

eleven internal splined plates (9) and ten externally splined plates (10).

Model 2855N tractors before serial number 654696 are originally equipped with a spacer (10A), spacer ring (13) that is 34.2 mm (1.35 inches) thick, eight internal splined plates (9) and seven externally splined plates (10).

Model 2855N tractors after serial number 654695 are originally equipped with a spacer ring (13) that is 34.2 mm (1.35 inches) thick, nine internal splined plates (9) and eight externally splined plates (10). Spacer (10A) is NOT used.

Model 2955 (not high-clearance) tractors before serial number 655825 are originally equipped with a spacer ring (13) that is 34.2 mm (1.35 inches)

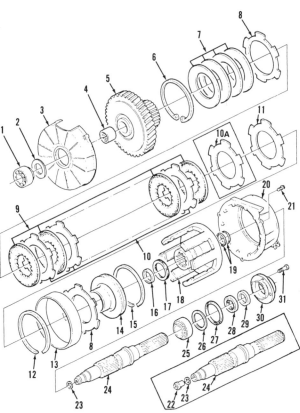

Fig. 23—Exploded view of front-wheel drive clutch used on some models. Refer to text for explanation of key differences.

1. Taper roller bearing	16. Thrust washer
2. Spacer	17. Seal ring
3. Cover (rear)	18. Clutch drum
4. Bushing	19. "O" rings
5. Clutch hub & gear	20. Cover (front)
6. Snap ring	21. Cap screw (3)
7. Belleville springs	22. Sleeve
8. Pressure rings	23. Seal ring
9. Internal splined discs	24. Clutch shaft
10. External splined discs	25. Front bearing cone & cup
10A. Spacer	26. Shim
11. Pressure ring	27. "O" ring
12. Snap ring	28. Oil seal
13. Spacer ring	29. Seal washer
14. Piston	30. Bearing quill
15. Piston ring	31. Cap screws

thick, nine internal splined plates (9) and eight externally splined plates (10).

Model 2955 (not high-clearance) tractors after serial number 655824 are originally equipped with a spacer ring (13) that is 37 mm (1.46 inches) thick, ten internal splined plates (9) and nine externally splined plates (10).

Model 2955 high-clearance tractors before serial number 646531 are originally equipped with a spacer ring (13) that is 34.2 mm (1.35 inches) thick, nine internal splined plates (9) and eight externally splined plates (10).

Model 2955 high-clearance tractors after serial number 646530 are originally equipped with a spacer ring (13) that is 38.3 mm (1.5 inches) thick, eleven internal splined plates (9) and ten externally splined plates (10).

On all models, align internal splines of clutch discs (9), position spacer ring (13) over clutch drum, then install rear pressure ring (8) with rounded edge down toward spacer (13). Install the four Belleville springs (7) as shown in Fig. 24. Temporarily install clutch hub and gear to align internal splined discs (9—Fig. 23), then use spring compressor JDT-24A or a press, compress Belleville springs and install snap ring (6). Install thrust washer (16) in clutch hub and gear (5) with chamfered side facing clutch drum. Carefully insert clutch hub and gear into clutch discs. If splines of clutch discs (9) are not aligned, it may be necessary to compress Belleville springs (7) again, remove snap ring (6), release springs and again align the splines. When the assembled clutch can be installed fully over hub, install covers (3 and 20) and secure with three cap screws (21).

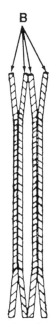

Fig. 24—Sectional view showing correct installation of front-wheel drive clutch Belleville springs (B) used.

Before installing clutch assembly, check and adjust, if necessary, clutch shaft axial free play. Install rear bearing cup (1) in clutch housing. Install clutch shaft (24) with rear bearing cone (1), spacer (2) and front bearing (25). Install bearing quill (30) without shims (26) and secure with three nuts (32). With a dial indicator against front end of clutch shaft, move shaft forward and rearward by hand and check axial end play with a dial indicator. Remove bearing quill (30) and add sufficient thickness of shims to reduce end play of shaft (24) in bearings to 0.02 mm (0.0008 inch) preload to 0.03 mm (0.0012 inch) play.

Remove rear bearing cone (1) and spacer (2) from shaft. Use light grease to stick bearing cone in bearing cup (1) in clutch housing. Install new seal ring (23) on sleeve (22) and new "O" rings (19) in grooves of shaft at each side of hydraulic oil port. Shaft (24) is different on later models and sleeve (22) is not used. Hold clutch assembly into clutch housing and insert clutch shaft until spacer (2) can be installed. Install spacer with chamfered side facing bearing (1). Push shaft fully to the rear and install front bearing cup (25). Install the determined shim pack (26) and new "O" ring (27). Install bearing quill (30), with seal washer (29) and oil seal (28) and secure with stud nuts or screws (31).

Reinstall remaining parts by reversing removal procedures. Fill transmission to level mark on dipstick. Install drive shaft as outlined in paragraph 7.

CLUTCH OIL PRESSURE TEST

All Models with MFWD

17. To check front-wheel drive clutch oil pressure, operate tractor until transmission oil temperature is 40° C (100° F). Remove test plug (5—Fig. 25) and install a 2000 kPa (300 psi) test gage.

> **CAUTION: If the plug has ⁹⁄₁₆ or ⁵⁄₈ inch head, the threads should be English; if plug removal requires ³⁄₄ inch or metric size wrench, port probably has metric threads. If there is any question, check thread size before attaching test gage.**

With engine operating at 1000 rpm and front-wheel drive control switch in "Disengaged" position, operating pressure should be approximately 1050 kPa (150 psi). With control switch in "Engaged" position, pressure gage should read zero. If not, repair solenoid as outlined in the following paragraph.

To check for internal leakage, such as caused by leaking clutch piston seals, install two (matched) 2000 kPa (300 psi) gages as described in Fig. 27. With transmission in "Neutral," dash "MFWD" switch OFF and engine operating at 1000 rpm, pressure at the two test points should be within 170 kPa (25 psi) of the same pressure. If special test cover No. 0745 is used instead of "T" fitting shown in Fig. 27, maximum allowable difference can be 185 kPa (27 psi).

SOLENOID

All Models with MFWD

18. When key switch is ON and front-wheel drive control switch is in "Disengaged" position or OFF position, the mechanical front-wheel drive solenoid is energized causing solenoid spool to direct pressurized oil to the clutch piston. Clutch piston compresses the Belleville springs (7—Fig. 26) disengaging the front-wheel drive clutch. When control switch is in "Engaged" position, there is no current to the solenoid and a spring forces spool to the OFF position. In this position, there is no oil pressure at clutch piston and the Belleville springs engage the clutch. The outside of the control solenoid will be magnetized when the key switch is ON and the MFWD switch is in "Disengaged" or OFF position. This magnetic field should be easily observed by passing a steel tool such as a screwdriver near the solenoid. Check condition of switch and remainder of electrical circuit if magnetic field is not observed.

To remove the solenoid, remove drain plug (1—Fig. 25) and drain oil. Unbolt and remove solenoid guard (8). Disconnect wires from solenoid. Remove "E" ring (10 or 22) and pull solenoid coil (11 or 25) from core (14 or 31). Remove Allen screws (12 or 27) and retaining plate (13 or 28). Valve parts can be pulled from bore in pan if cleaning, repair or renewal is required. Early and late parts are different, but operation and service procedures are similar.

Clean and inspect all parts for wear or damage and renew as necessary. Spool and sleeve are available only as a matched set. Use new back-up washer and "O" rings when reassembling and reinstall by reversing the disassembly and removal procedures.

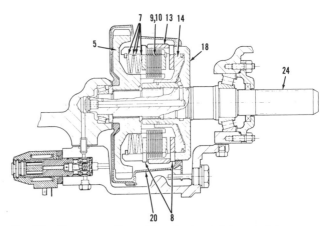

Fig. 26—Cross section of front-wheel drive clutch, cover and valve. Differences may be noted between model shown and some models. Refer to text.

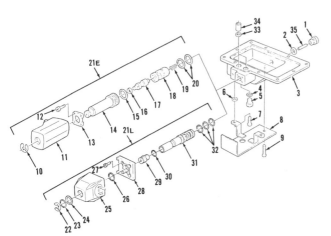

Fig. 25—Exploded view of clutch solenoid and relative parts. Later solenoid/valve assembly is shown at 21L and early type is at 21E.

1. Drain plug	19. Spring
2. Seal ring	20. "O" rings
3. Oil pan	21E. Solenoid/valve unit
4. "O" ring	21L. Solenoid/valve unit
5. Test plug	22. "E" ring
6. Washer	23. "O" ring
7. Cap screw (2)	24. "O" ring
8. Solenoid guard	25. Solenoid coil
9. Cap screw	26. "O" ring
10. "E" ring	27. Allen screw (4)
11. Solenoid coil	28. Retaining plate
12. Allen screw (4)	29. Plug
13. Retaining plate	30. "O" ring
14. Solenoid core	31. Core sleeve
15. "O" ring	32. "O" rings
16. Back-up washer	33. Seal rings (2)
17. Spool	34. Tube
18. Sleeve	35. Pin (8 x 36 mm)

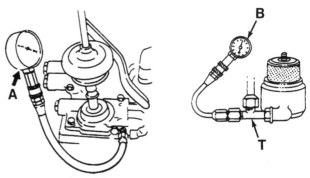

Fig. 27—Two gages should be installed at locations shown to check for internal clutch leakage on models with MFWD. Gage (A) is attached to port shown in clutch control valve housing and gage (B) attached at filter. The tee (T) or the special ported filter cover must be installed to attach gage (B). Since difference in pressure at the two ports is measured, it is important that the gages are matched to indicate exactly the same pressure readings.

POWER STEERING SYSTEM

A closed center-type power steering system is used on some 2755 and 2855N two-wheel drive models and is identified by a mechanical connection from steering shaft arm (12—Fig. 33) to the steering Bellcrank (13—Fig. 2 or Fig. 3) via a drag link. Pressurized oil is furnished by the main hydraulic pump via a priority-type pressure control valve that is bolted to lower right side of transmission housing. In the event of hydraulic failure or engine stoppage, steering can still be accomplished with the mechanical advantage built into steering valve assembly. Refer to paragraph 25 and following for service to the hydrostatic steering system used on other models.

TROUBLE-SHOOTING

2755 and 2855N Models So Equipped

19. Problems that develop in the power steering system may appear as sluggish steering, loss of power steering, power steering in one direction only or excessive noise in the power steering unit.

1. Sluggish steering usually can be attributed to:
   a. Leakage past valve seats that usually produces slow steering in one direction only.
   b. Piston sealing ring or "O" ring failure.
   c. Steering valve body (or bodies) leaking.

2. Loss of power steering usually can be attributed to:
   a. Insufficient oil supply from transmission oil pump or main pump.
   b. Pressure control valve out of adjustment.
   c. Clogged hydraulic oil filter.

Chattering system components are usually the result of "O" ring failure in unit or air leak on suction side of pump.

Because the power steering valve is a complete self-contained unit, its only external requirement for proper operation is adequate hydraulic pressure. The transmission pump serves as a charging pump for the main hydraulic pump, which, in turn, supplies the steering valve via the pressure control valve. The priority-type pressure control valve supplies oil to the steering valve first; then to other hydraulic functions.

Pressure control valve output should be tested as outlined in paragraph 22, main hydraulic pump tested as in paragraph 153 and transmission pump tested as in paragraph 120 or 127.

Raise front wheels off the ground and with engine stopped, turn steering wheel from one extreme turn position to the other. No binding or hard spots should be encountered. Lower front wheels to the ground, start engine and turn steering wheel from stop to stop. Steering effort should be equal in both directions. Excessive steering effort in one direction only would indicate trouble in steering valve body and/or steering valve for that direction. Excessive steering effort in both directions might indicate trouble with piston "O" ring or seal ring, but may also be caused by excessive weight on front wheels. Internal leakage may be caused by dirty or worn valve seats in valve bodies (14—Fig. 28), or by a valve body shim pack incorrectly adjusted . If adjuster oil seal (2) is leaking, return oil passage may be obstructed.

Fig. 28—Exploded view of typical power steering valve. Item (35) is shown in Fig. 33.

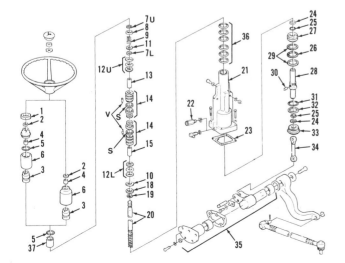

S. Shims	19. Snap ring
V. Valve sleeves	20. Steering wheel shaft
1. Jam nut	21. Housing
2. Oil seal	22. Inlet check valve
3. Adjuster	23. Gasket
4. Bushing	24. Back-up rings
5. "O" rings	25. "O" rings
6. Sleeve	26. Seal ring
7L. Snap rings	27. Piston
7U. Snap rings	28. Piston rod
8. Special washer	29. Back-up rings
9. Spring	30. Pin
10. Shim	31. "O" ring
11. Thrust washer	32. Back-up ring
12L. Lower thrust bearing	33. Piston rod guide
12U. Upper thrust bearing	34. Steering rod
13. Sleeve	35. Steering shaft
14. Valve bodies	36. "O" rings
15. Sleeve	37. Spacer
18. Special washer	

STEERING VALVE

2755 and 2855N Models So Equipped

20. REMOVE AND REINSTALL. Remove steering wheel and interfering covers from around steering valve assembly. Disconnect hydraulic inlet pressure line and remove cover (3—Fig. 33). It may be necessary to detach speed control linkage from cover of some models. Center steering shaft yoke in cover hole by turning steering, then remove cap screw (7), lock washer (8) and pin retainer (9). Thread a 3/8 inch cap screw into end of pin (6) and remove pin. Steering valve now can be disconnected from dash and clutch housing, and removed as a unit.

Reinstall in reverse order of removal. Coat threads of screw (7) with Loctite 270 and tighten to 15 N•m (11 ft.-lbs.) torque. Start engine and cycle system several times to remove air that may be in system.

21. OVERHAUL. With steering valve removed as outlined in paragraph 20, refer to Fig. 28 and proceed as follows: Remove jam nut (1) and oil seal (2). Use special JDH 41-1B wrench or equivalent to remove adjuster (3), then slide sleeve (6) from housing. Rotate shaft (20) counterclockwise until piston rod (28) is completely free of steering housing. Withdraw steering shaft (20) and steering valve from housing (21). Compress special washer (8) against spring (9), then remove top snap ring (7). Release spring and slide valve parts from shaft.

> NOTE: Upper and lower valve body parts (14) must not be intermixed. If renewal is necessary, renew as a preadjusted assembly. Special tool (JDG183) is required to determine the correct thickness of shims (S—Fig. 29). Also, lower valve housing has two additional oil passages (Fig. 30).

Use a special pronged wrench to remove piston (27—Fig. 28) from piston rod (28). Pin (30) must be pressed from piston rod for removal of steering rod (34).

Renew adjuster oil seal (2) and bottom seal in bore. Bushing in adjuster (3) should be renewed if worn excessively. Bushing should be installed until flush with bottom of seal bore in adjuster (3). Inspect all parts for distortion, wear or rough spots on threads and ring lands and renew as necessary.

Install piston (27) on piston rod (28) with dowel holes in piston on top. Use a special pronged wrench and tighten piston to 340 N•m (250 ft.-lbs.) torque. Assemble all parts on steering wheel shaft (20), beginning with snap ring (19) in groove. Be sure large chamfers on thrust bearing races (12L and 12U) are toward valve bodies. Clearance between snap ring (7L) and thrust bearing (12U) should be less than 0.1 mm (0.004 inch). If excessive, add shims at (10).

Install washer (11), spring (9) and special washer (8). Compress spring and install upper snap ring (7U). Renew all "O" rings, back-up rings and seal rings. Lubricate all parts before assembling. Assemble steering rod (34) into piston rod (28), small end first, and press pin (30) until flush with outside surface of piston rod. Slip piston rod guide (33) onto piston rod and install assembly into housing. Insert assembly on

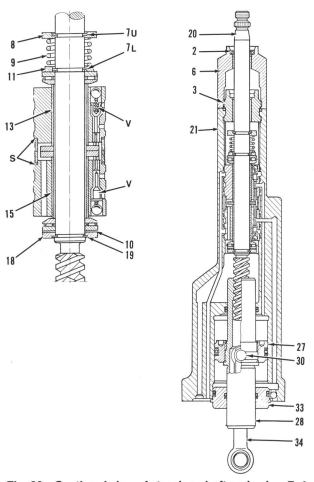

Fig. 29—Sectional view of steering shaft and valve. Refer to Fig. 28 for legend.

Fig. 30—View showing the two additional oil return passages in the lower valve housing.

steering wheel shaft and parts into housing and thread shaft into piston rod. Install sleeve (6) and wrap threads of steering wheel shaft with plastic tape to protect lips of seal (2) as adjuster (3) is screwed into housing. Tighten adjuster to 100 N·m (75 ft.-lbs.) torque. Tighten jam nut (1) to 40 N·m (30 ft.-lbs.) torque.

Reinstall unit on tractor with front wheels in straight ahead position. Install steering rod pin (6—Fig. 33) through yoke (10). Install pin retainer (9), lock washer (8) and cap screw (7). Coat threads of screw (7) with Loctite 270 and tighten to 15 N·m (11 ft.-lbs.) torque. Install cover (3) with new gasket. Install steering wheel with spoke pointing straight down and tighten nut to 70 N·m (50 ft.-lbs.) torque. Start engine and cycle system several times to purge air from steering valve.

PRESSURE CONTROL VALVE

2755 and 2855N Models So Equipped

22. PRESSURE TEST. Remove the ⅜-inch plug from right side of hydraulic pump, then attach a 2000 kPa (3000 psi) test gage as shown in Fig. 32. Start engine and operate at approximately 1900 rpm. Check lift system operations and pump standby pressure, which should be 15,900-16,200 kPa (2300-2350 psi). Readjust pump stroke control valve (see paragraph 155) to 10,345 kPa (1500 psi), then attempt to operate rockshaft. The lift system should not function, and if it does, pressure control valve is faulty and should be repaired as in paragraph 23 before proceeding further.

To continue testing pressure control valve, completely lower rockshaft and place rockshaft control lever in raise position. With engine running at 1900 rpm, adjust main hydraulic pump stroke control

(raise pressure) until rockshaft raises at its normal rate. This is the regulating point of the pressure control valve and gage should read 11,725-12,410 kPa (1700-1800 psi). If pressure is not as stated, disconnect front oil line, remove fitting (11—Fig. 31) and vary shims (7) as required. Shims are 0.76 mm (0.030 inch) thick and one shim will change pressure 250-280 kPa (35-40 psi).

If pressure control valve cannot be adjusted satisfactorily, remove and service valve as outlined in paragraph 23. Readjust main hydraulic pump stroke control valve to 15,900-16,200 kPa (2300-2350 psi) standby pressure. Remove test gage and reinstall plug.

23. R&R AND OVERHAUL. To remove pressure control valve, drain transmission. Disconnect inlet (front) oil line and, if equipped with remote hydraulics, the rear (outlet) oil line. Unbolt and remove unit from transmission housing.

With valve removed, remove fitting (11—Fig. 31), then remove spool (9), shims (7) and spring (6) from housing (3). Retain shims for subsequent installation.

Inspect spool and housing for wear, scoring or other damage. Spring has a free length of 117 mm (4.62 inches) and should test 200-250 N (45-55 lbs.) when compressed to a length of 89 mm (3.5 inches).

When reinstalling, attach oil lines before final tightening of mounting bolts.

STEERING CROSS SHAFT

2755 and 2855N Models So Equipped

24. REMOVE AND REINSTALL. Steering cross shaft (10—Fig. 33) can be removed without removing steering valve and steering wheel assembly as follows:

Drain oil from steering shaft housing by removing plug on right side of clutch housing. Remove cover (3), center yoke of steering shaft in cover opening and remove cap screw (7), lock washer (8) and pin retainer (9). Thread a cap screw into pin (6) and remove pin. Take out button plug on left side of clutch housing for

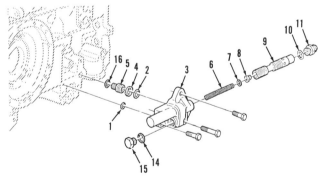

Fig. 31—Exploded view of pressure control valve. Orifice (8) is integral with valve (9).

1. "O" ring	8. Orifice
2. "O" ring	9. Control valve
3. Valve housing	10. "O" ring
4. Back-up ring	11. Connector
5. Fitting	14. "O" ring
6. Spring	15. Plug or fitting
7. Shim	16. "O" ring

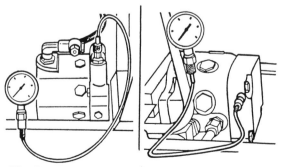

Fig. 32—Connect a 20,000 kPa (3000 psi) test gage to main pump as shown for pressure control valve test.

access to cap screw (14). Refer to Fig. 34 and remove cap screw and steering shaft arm. Inset in Fig. 34 shows cover removed for access to steering shaft yoke, pin retainer and pin. Turn steering wheel full right, make sure steering shaft yoke is aligned with right cover opening and bump steering shaft out right side of clutch housing. Inspect bushing (4—Fig. 33) in cover (3) and bushing in clutch housing and renew if necessary.

Install by reversing the removal procedure. Coat threads of screw (7) with Loctite 270 and tighten to

15 N·m (11 ft.-lbs.) torque. After cover (3) is installed, tighten arm attaching screw (14) securely, strike arm with a hammer, then tighten cap screw to 325 N·m (240 ft.-lbs.) torque. Repeat procedure until screw remains at same (correct) torque after striking to be sure that arm is fully seated.

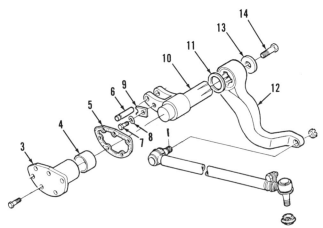

Fig. 33—View of steering shaft component parts that are contained in top of clutch housing of models without hydrostatic steering.

3. Cover
4. Bushing
5. Gasket
6. Pin
7. Cap screw
8. Lock washer
9. Pin retainer
10. Steering shaft
11. Oil seal
12. Steering shaft arm
13. Special washer
14. Cap screw

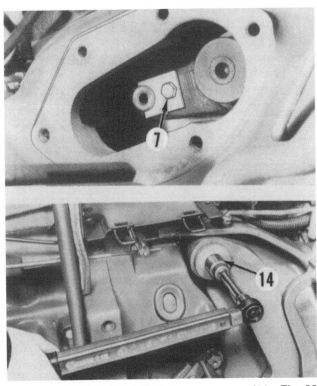

Fig. 34—Steering arm attaching cap screw (14—Fig. 33) is located on left side. Top view is of right side with cover removed showing screw (7).

HYDROSTATIC STEERING SYSTEM

A hydrostatic steering system is used on some models and consists of a steering valve assembly, safety valve block, pressure control valve and one or two steering cylinders. In the event of hydraulic failure or engine stoppage, manual steering can be accomplished by means of the geroter pump in the steering valve. Refer to paragraph 19 and following for power steering system used on other models.

TROUBLE-SHOOTING

All Models with Hydrostatic Steering

25. Some problems that may occur during operation of power steering and their possible causes are as follows:

1. Steering wheel hard to turn. Could be caused by:
   a. Defective hydraulic pump.
   b. Leaking or missing recirculating ball.
   c. Binding of mechanical parts in front steering system.
   d. Damaged ball bearings in steering column.
   e. Leaking steering cylinder.
   f. Binding valve spool and sleeve.

2. Steering wheel turns on its own. Could be caused by:
   a. Weak or broken leaf springs.

3. Steering wheel does not return to neutral position. Could be caused by:
   a. Jammed valve spool and sleeve.
   b. Leaking between valve sleeve and housing.

c. Dirt or metal chips between valve spool and sleeve.

4. Excessive steering wheel play. Could be caused by:
   a. Worn inner teeth of rotor.
   b. Worn upper flange of drive shaft.
   c. Weak or broken leaf springs.
   d. Worn drive shaft teeth .

5. Steering wheel rotates at steering cylinder stops. Could be caused by:
   a. Excessive leakage in steering cylinder(s).
   b. Excessively worn rotor and stator .
   c. Excessive leakage between valve spool and sleeve and between sleeve and housing.

6. Steering wheel "kicks" violently. Could be caused by:
   a. Incorrect adjustment between drive shaft and rotor.

7. Steering wheel responds too slowly. Could be caused by:
   a. Not enough oil.
   b. Worn steering control valve.

8. Tractor steers in wrong direction. Could be caused by:
   a. Incorrectly connected steering cylinder hoses.
   b. Incorrect timing of drive shaft to rotor.

STEERING VALVE

All Models with Hydrostatic Steering

26. STEERING VALVE LEAKAGE TEST. Operate tractor to warm hydraulic oil to 60°-70° C (140°-160° F). Cycle steering valve lock to lock approximately 15 times to stabilize temperature. Detach both steering cylinder lines from the safety valve ports at the side of control valve. Install caps on lines to prevent entrance of dirt and plug the two steering valve ports with high-pressure metal plugs to prevent leakage of hydraulic fluid during test. Operate steering wheel to the left and apply 7 N•m (60 in.-lbs.) torque to steering shaft. Continue turning steering to the left and check number of turns attained in one minute. Repeat this test turning steering to the right. If steering wheel turns more than 1.5 revolutions in one minute, excessive leakage is indicated. Remove steering valve and repair as outlined in the following paragraphs.

27. REMOVE AND REINSTALL. On tractors without Sound Gard Body, remove steering wheel. Swing both dash side panels upward. Identify and disconnect hydraulic lines from safety valve block on steering valve. Plug or cap all openings immediately to prevent entrance of dirt. Unbolt and remove steering column cover. Remove four retaining cap screws and remove steering valve and column assembly from below dash.

On tractors equipped with Sound Gard body, disconnect battery ground straps. Push steering column fully forward, then remove steering wheel cover and the release wheel. Remove steering wheel nut and use puller to remove steering wheel. Remove snap ring and slide steering column extension from upper steering shaft. Remove cover below dash, then unbolt and remove dash side covers. Disconnect all electrical plugs at dash and disconnect shut-off cable. Unbolt and remove dash from panel bracket. Identify and disconnect hydraulic lines from safety valve block on steering valve. Cap or plug all openings immediately to prevent dirt from entering system. Remove steering tilt lever, then unbolt and remove steering column and steering valve assembly.

On all tractors, reinstall steering valve assembly by reversing removal procedure. Start engine and operate steering system lock to lock several times to purge air from system.

28. OVERHAUL. Remove steering valve as described in paragraph 27, then thoroughly clean exterior of unit. Unbolt and remove steering column, then place steering valve in a special holding fixture. Remove the two Allen screws and separate safety block (40 or 41—Fig. 36) with intermediate plate (42) and "O" rings (43) from valve of models so equipped. On all models, remove the seven cover retaining screws (10—Fig. 35 or Fig. 36). Remove cover (9), stator (7), rotor (5), spacer ring (2) and "O" rings (6 and 8). Remove distributor plate (4), drive shaft (1) and "O" ring (3). Suction valve parts (37, 38 and 39—Fig. 35) will fall from threaded holes of models so equipped. On all models, hold steering valve vertically and turn valve spool and sleeve to align cross pin (20—Fig. 35 or Fig. 36) parallel to flat side of housing. With cross pin in this position and housing in horizontal position, remove sleeve (22), spool (21), thrust bearing (18) and bearing races (17 and 19) from housing. Remove cross pin (20) from rotary valve and separate spool (21) from sleeve (22). Remove leaf springs (23) from spool. Remove snap ring (12), gland bushing and seal (13) with oil seal (11), "O" ring (14) and quad seal (16—Fig. 36) or seal (30, 31 and 32—Fig. 35) from housing. Unscrew retainer (24—Fig. 35 or Fig. 36) and remove valve seat (26) with "O" rings (25 and 27), check ball (28) and ball stop (29). Safety block valve (33 to 36—Fig. 35) can be removed after unscrewing plug (34) of models so equipped; however, pressure must be checked and adjusted as outlined in paragraph 29 if plug (34) is turned, thus changing adjusted pressure.

Clean and inspect all parts for excessive wear or other damage and renew parts as necessary. Housing

(15—Fig. 35 or Fig. 36), spool (21) and sleeve (22) are available only as an assembly. Use all new "O" rings and seals when reassembling. Lubricate all interior parts with clean steering fluid.

Insert spool (21) into sleeve (22) aligning leaf spring slots. Install leaf springs (23), in two sets of two with two flat springs in the middle as shown at (23 and 23S—Fig. 35) of models so equipped or in two sets of three arch to arch as shown at (23—Fig. 36). Use special tool KML 10018-3 when installing leaf springs on all models. Insert cross pin (20—Fig. 35 or Fig. 36) into sleeve and spool. Install "O" ring (14), gland bushing seal (13) with oil seal (11) and snap ring (12). Place bearing race (19), thrust bearing (18), bearing race (17) and quad ring (16—Fig. 36) or seal parts (30, 31 and 32—Fig. 35) on spool. Insert ball stop (29—Fig. 35 or Fig. 36), check ball (28), valve seat (26) with new "O" rings (25 and 27) and retainer (24). Tighten

retainer to a torque of 12 N·m (100 in.-lbs.). Lubricate sleeve and install spool and sleeve assembly.

The balance of reassembly is the reverse of disassembly procedure, keeping the following points in mind: When installing rotor (5), make certain that timing dot on end of drive shaft is aligned with a valley on outer side of rotor. Tighten end cover retaining cap screws (10) to an initial torque of 10-15 N·m (7-10 ft.-lbs.), then to the final torque of 25-30 N·m (18-23 ft.-lbs.). Be sure that suction valve parts (37, 38 and 39—Fig. 35) are reinstalled in proper holes before installing screws (10). Tighten Allen screws attaching external safety valve block (40 or 41—Fig. 36) to 70 N·m (50 ft.-lbs.) torque. Hose connectors should be tightened to 50 N·m (35 ft.-lbs.) torque on all models. If setting of plug (34—Fig. 35) has been disturbed, refer to paragraph 29 for testing and adjusting proper opening pressure.

SAFETY VALVE BLOCK

All Models with Hydrostatic Steering

A safety shock valve is used to act as a relief valve should a front wheel strike a solid object. Piston in steering cylinder is forced to one side and oil pressure

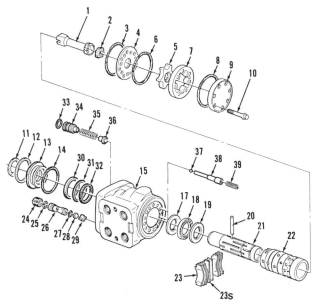

Fig. 35—Exploded view of Char-Lynn hydrostatic steering valve with integral safety shock valves (33 through 36). Notice that the two center springs (23S) are straight.

1. Drive shaft
2. Spacer
3. "O" ring
4. Distributor plate
5. Rotor
6. "O" ring
7. Stator
8. "O" ring
9. End cover
10. Cap screw (7)
11. Oil seal
12. Snap ring
13. Gland bushing seal
14. "O" ring
15. Housing
17. Bearing race
18. Thrust bearing
19. Bearing race
20. Cross pin
21. Valve spool

22. Valve sleeve
23. Leaf springs (4)
23S. Flat leaf springs (2)
24. Retainer
25. "O" ring
26. Valve seat
27. "O" ring
28. Check ball
29. Ball stop
30. Back-up ring
31. "O" ring
32. Seal
33. "O" ring
34. Adjusting plug
35. Spring
36. Shock valve
37. Suction valve ball
38. Suction valve pin
39. Spring

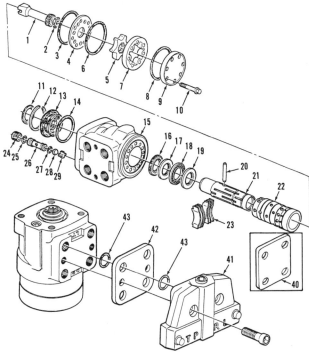

Fig. 36—Exploded view of Char-Lynn hydrostatic steering valve. Notice that all six leaf springs (23) are arched on model shown. If safety shock valve and suction valve are not located in valve housing as shown in Fig. 35, similar valves are located in separate valve block (40 or 41) as shown in Fig. 37 or Fig. 38. Refer to Fig. 35 for legend except the following.

16. Quad ring
23. Leaf springs (6)
40. Safety valve block (Fig. 37)
41. Safety valve block (Fig. 38)
42. Plate
43. "O" rings

increases greatly in hydraulic steering lines. The suction valve on pressure-loaded side of safety valve closes immediately. The safety shock valves and suction valves may be located in the steering valve housing as shown at (33 to 39—Fig. 35), or in a separate safety valve block (40 or 41—Fig. 36). The specific type used will depend upon model, serial number range and equipment installed on the specific tractor. The safety valve should open at 20,500 kPa (2980 psi) on 2750 models, 21,000 kPa (3050 psi) on other models. When the safety valve opens, oil is allowed to flow to the low-pressure side of steering cylinder. The check valve prevents oil from returning to hydraulic pump when pump is not operating.

29. TEST AND ADJUST. To test and adjust safety valve block, first remove steering valve assembly as outlined in paragraph 27. Connect a manual hydraulic pump with a 0-34,500 kPa (0-5000 psi) pressure gage to cylinder line port on steering valve (15—Fig. 35) or to appropriate port (3 or 4—Fig. 37 or Fig. 38) of valve block body. Connect a line to return flow port (1) and insert other end into a suitable container. Plug remaining two ports with M18 × 1.5 × 15 mm cap screws. Operate manual pump until a pressure of about 20,500 kPa (2980 psi) is obtained, then slowly increase pressure until valve opens. Pressure will drop suddenly when valve opens. The safety valve should open at 20,500 kPa (2980 psi) on 2750 models, 21,000 kPa (3050 psi) on other models.

If necessary, adjust safety valve as follows: Remove end cap and turn adjusting nut (6—Fig. 37 or Fig. 38) or adjusting plug (34—Fig. 35) clockwise to increase pressure or counterclockwise to decrease pressure until correct pressure is obtained.

With pressure correctly adjusted, reinstall steering valve as in paragraph 27.

PRESSURE CONTROL VALVE

All Models with Hydrostatic Steering

A priority-type pressure (flow) control valve is located at lower right side of transmission. This valve assures oil supply at sufficient pressure to hydrostatic steering at all times when engine is operating.

30. R&R AND OVERHAUL. To remove the pressure control valve, disconnect hydraulic pump pressure line and line to selective control valve at pressure control valve. Unbolt and remove pressure control valve from transmission case. Remove adapter (14—Fig. 39) with "O" ring (13), then withdraw valve spool (12), orifice (11), shims (10) and spring (9).

Check valve spool (12) and housing (8) for scoring or other damage. Spring free length should be 117 mm (4.62 inches). Spring should test 200-250 N (45-55 lbs.) when compressed to a length of 89 mm (3.5 inches). Check orifice (11) for wear and renew as necessary.

When reassembling, use same number of shims that were removed. Reassemble and reinstall pressure control valve. Check and adjust pressure if necessary as outlined in the following paragraph.

31. PRESSURE TEST AND ADJUST. To check pressure control valve, first drain auxiliary reservoir. Remove upper right plug on right side of main pump. Install a 35,000 kPa (5000 psi) test gage in pump. Operate tractor until hydraulic oil is approximately 66° C (150° F). Install a jumper hose across left selective control valve. Set metering valve at maximum and pull left control valve to circulate oil. With

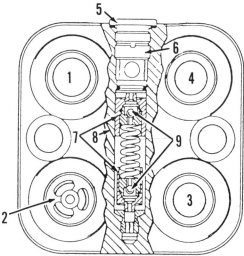

Fig. 37—Cutaway view of safety valve block used on some models equipped with Sound Gard Body. On some models, safety shock valve and suction valves may be located in steering valve body as shown at (33 to 39—Fig. 35) or may use external valve block as shown in Fig. 38.

1. Return flow port
2. Inlet port w/check valve
3. Cylinder line port
4. Cylinder line port
5. End cap
6. Adjusting nut
7. Safety valve
8. Return flow passage
9. Suction valves

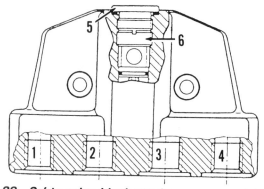

Fig. 38—Safety valve block used on some models not equipped with steering valve with integral safety and suction valves as shown at (33 to 39—Fig. 35). Refer to Fig. 37 for legend.

engine operating at 800 rpm, test gage should read 11,600-12,300 kPa (1700-1800 psi). If not, stop engine, disconnect pump pressure line and remove adapter (14—Fig. 39). Withdraw valve spool (12) and orifice (11), then add or remove shims (10) as necessary. Adding one shim will increase pressure 250-280 kPa (35-40 psi). Reassemble and retest pressure.

2855N Models

31A. A non-repairable pressure reduction valve is used on 2855N models. To test the unit, refer to Fig. 40 and install a suitable pressure gage as shown. Parts (A, B, C, D and E) are included in pressure test kit JT0570. Pressure reduction valve should reduce pressure from 16,000 kPa (2320 psi) from the hydraulic pump to 12,000 kPa (1740 psi) for the hydrostatic steering. Pressure is not adjustable and unit should not be disassembled.

STEERING CYLINDERS

Various types of steering cylinders are used on models with hydrostatic steering. Refer to appropriate following paragraphs for overhaul procedures.

2755 GP Models

32. GP tractors with fixed tread front axle (Fig. 1) and hydrostatic steering are equipped with two double-acting steering cylinders as shown in Fig. 41. Cylinders can be removed after disconnecting hoses and removing cotter pins and retaining pins. Cover openings in cylinder to prevent entrance of dirt and plug hoses to reduce leakage and to prevent entry of dirt. To disassemble, remove set screw, then unscrew rod guide (37). Nut (30) is self-locking. "O" ring (35)

is installed in seal groove under seal ring (36). Heat seal ring (36) in oil to approximately 55° C (130 ° F) before installing in groove. Tighten nut (30) to 350 N•m (260 ft.-lbs.) torque.

Models with Two Single-Acting Cylinders

33. On tractors without front-wheel drive and with two single-acting steering cylinders, remove either steering cylinder as follows: Disconnect hydraulic line at steering cylinder. Plug or cap openings to prevent dirt from entering system. Support front of tractor and remove front wheel. Disconnect tie rod

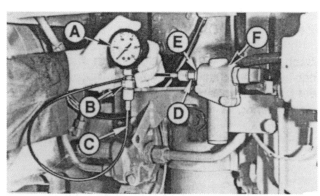

Fig. 40—View of pressure gage connections used to check the pressure reduction valve (F) used on 2855N models. Valve cannot be serviced and new valve must be installed if pressure is incorrect.

A. Pressure gage (FKM10207)
B. Adapter (FKM10302)
C. High pressure hose (FKM10209)
D. Adapter (FKM10303)
E. Male fitting (FKM10305)
F. Pressure reduction valve

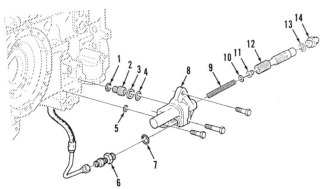

Fig. 39—Exploded view of pressure (flow) control valve used.

1. "O" ring	8. Valve housing
2. Adapter	9. Spring
3. Back-up ring	10. Shim
4. "O" ring	11. Orifice
5. "O" ring	12. Valve spool
6. Adapter	13. "O" ring
7. "O" ring	14. Adapter

Fig. 41—Exploded view of double-acting steering cylinder used on GP models with axle shown in Fig. 1.

1. Grease fitting	35. "O" ring
28. Cylinder	36. Back-up ring
29. Bushings	37. Piston rod guide
30. Nut	38. Piston rod seal
31. Wear ring	39. Wiper seal
32. "O" ring	40. Piston rod
33. Sealing ring	41. Bushing
34. Piston	

from steering arm. Unbolt and remove the axle extension from the axle main member. Detach cylinder ball joint end, unpin rod end and remove cylinder.

To disassemble cylinder, first clean exterior of cylinder and lightly clamp cylinder barrel in a vise. Using a screwdriver through hydraulic port, force snap ring out of groove and off end of piston rod as shown in Fig. 43. Pull piston rod (7—Fig. 42) from cylinder barrel (6). Remove wiper seal (1) and piston rod seal (2) from barrel.

Clean and inspect all parts for excessive wear or other damage and renew as necessary. Install new piston rod seal (2) with lips of seal facing inward, then install new wiper seal (1). Position snap ring (3) in bottom of cylinder barrel. Slide piston rod into cylinder barrel and use screwdriver to install snap ring into groove on piston rod.

Reinstall cylinder by reversing removal procedure. Tighten cylinder retaining slotted nut to a torque of 100 N·m (70 ft.-lbs.) and lock with cotter pin. Tighten axle extension bolts to a torque of 400 N·m (300 ft.-lbs.).

2855N Models with MFWD

34. REMOVE AND REINSTALL. The double-acting steering cylinder used on 2855N models with MFWD can be removed as follows: Disconnect hydraulic lines from steering cylinder and cover all openings to prevent entrance of dirt. Remove cotter pins and slotted nuts from tie rod ends at both steering knuckle arms, then remove cotter pin and remove the slotted nut (13—Fig. 44) from cylinder mounting ball joint (15). Turn steering to the end of the right

hand stop, detach tie rod ends and remove cylinder assembly. The cylinder mounting ball joint (15) can be removed after removing screw (19—Fig. 46).

Install new assembly if cylinder mounting ball joint is worn excessively. Coat plastic bushing (16) with petroleum jelly when installing. Adjust axial play of cylinder mounting ball joint to 0.2-0.4 mm (0.008-0.016 inch) by varying the thickness of shims (17). Tighten cap screw (19) to 120 N·m (85 ft.-lbs.) torque. Thread tie rod ends onto steering rod as necessary to provide 0-3 mm (0-⅛ inch) toe-in, then tighten clamps (5) to 30 N·m (25 ft.-lbs.) torque. Tighten slotted nut (13—Fig. 44) to 210 N·m (155 ft.-lbs.) torque and tie rod end slotted nuts (2) to 95 N·m (70 ft.-lbs.) torque, then install cotter pins (1 and 14).

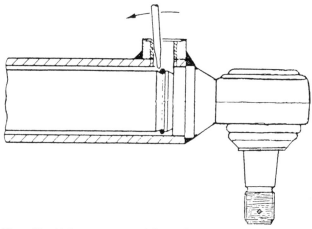

Fig. 43—Using a screwdriver through hydraulic port, force snap ring out of groove and off end of piston rod.

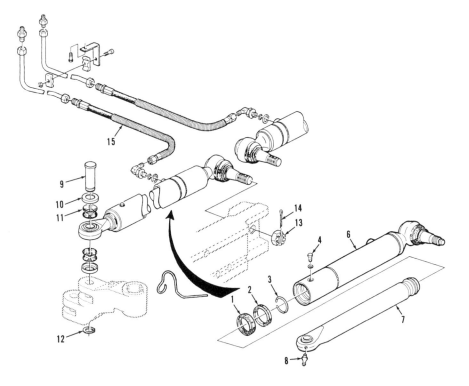

Fig. 42—Exploded view typical of single-acting steering cylinder used on models with adjustable front axle and hydrostatic steering. Two cylinders are used.

1. Wiper seal
2. Piston rod seal
3. Snap ring
4. Bleed plug
6. Cylinder barrel
7. Piston rod
8. Grease fitting
9. Pin
10. Washer
11. Seal
12. Snap ring
13. Slotted nut
14. Cotter pin
15. Hoses

35. OVERHAUL. Loosen clamps (5—Fig. 44) and remove both tie rod ends (4 and 28). Note that one end has left hand thread. Clean the exterior of the steering cylinder, then remove external snap rings (6). Push the piston rod guides (10) into cylinder far enough to remove the internal snap rings (9). The piston (24) and piston rod (27) can be used to push rod guides from cylinder bore after the internal snap rings are removed. To remove piston (24), remove snap ring (20) and special washer (21). Remove split rings (22), then pull piston from piston rod.

NOTE: Split rings (22) are available in various thicknesses. Be sure to keep sets of same thickness together.

New "O" ring (23) should be installed when piston is removed or unit is resealed.

Lubricate "O" ring (23) and position in center groove of rod (27). Install one set of split rings (22) in groove of rod. Slide piston (24) over the installed and lubricated "O" ring with flat side of piston toward the installed split rings.

NOTE: Split rings are available in various thicknesses, but it is important that each half is the same thickness.

Slide the piston tight against the installed split rings, then install the thickest possible split ring set in the remaining groove. Install special washer (21) with flat side toward piston and install snap ring (20). Back-up rings (25) should be on either side of piston sealing "O" ring (26).

Refer to Fig. 45 and install new wiper seals (7), oil seals (8) and "O" rings (11) in piston rod guides (10). Slide the piston and rod assembly into the lubricated bore of cylinder (12). Position one of the rod guides over the piston rod and into the cylinder bore. Push rod guide into bore far enough to install internal snap ring (9). Use the piston and rod to bump the installed rod guide out against the installed snap ring (9), then install the external snap ring (6). Install the remaining rod guide in a similar manner. Install the tie rod ends and install the unit as described in paragraph 34. Thread tie rod ends (4 and 28—Fig. 44) onto steering rod as necessary to provide 0-3 mm (0-⅛ inch) toe-in, then tighten clamps (5) to 30 N·m (25 ft.-lbs.) torque.

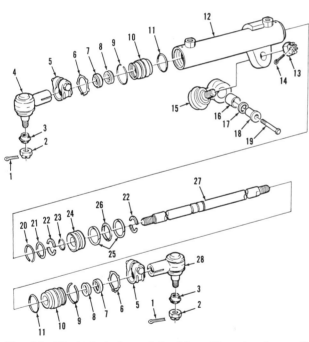

Fig. 44—Exploded view of double-acting steering cylinder used on 2855N models with MFWD.

1. Cotter pin	15. Ball joint
2. Slotted nut	16. Bushing
3. Rubber sleeve	17. Washer
4. Tie rod end	18. Special bushing
5. Clamp	19. Cap screw
6. Snap ring	20. Snap ring
7. Wiper seal	21. Special washer
8. Seal ring	22. Split ring
9. Snap ring	23. "O" ring
10. Piston rod guide	24. Piston
11. "O" ring	25. Back-up ring
12. Cylinder	26. "O" ring
13. Slotted nut	27. Piston rod
14. Cotter pin	28. Ball joint

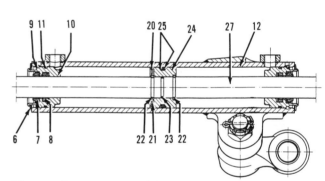

Fig. 45—Cross section of double-acting steering cylinder used on 2855N models with MFWD. Refer to Fig. 44 for legend.

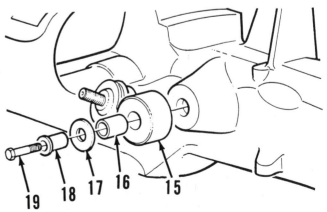

Fig. 46—Axial play of ball joint can be adjusted by varying thickness of washer (17). Refer to Fig. 44 for legend.

2750, 2755 and 2955 Models with MFWD

36. The double-acting steering cylinder used on these models must be disassembled for removal. Housing containing the steering cylinder sleeve is an integral part of front drive axle housing.

To disassemble steering cylinder, first disconnect tie rod ends from steering arms. Mark position of steering stop clamps (22—Fig. 47), then unbolt and remove stop clamps from each end. Loosen clamps (27) and unscrew left and right tie rods (28).

NOTE: Tie rod joints (26) are locked with Loctite and heat must be applied to area before unscrewing from piston rod (1).

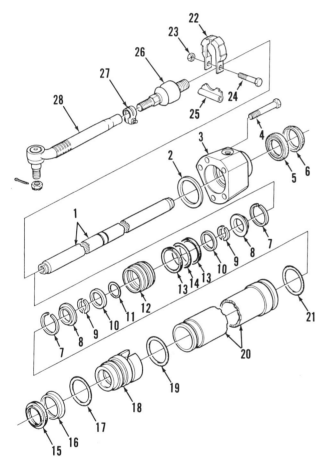

Fig. 47—Exploded view of double-acting steering cylinder components used on 2750, 2755 and 2955 four-wheel drive tractors.

1. Piston rod	15. Wiper seal
2. Shim	16. Piston rod seal
3. Outer guide	17. "O" ring
4. Cap screw (5)	18. Inner guide
5. Piston rod seal	19. "O" ring
6. Wiper seal	20. Sleeve
7. Snap rings	21. "O" ring
8. Washers	22. Stop
9. Split rings	23. Nut
10. Washers	24. Screw
11. "O" ring	25. Spacer
12. Piston	26. Tie rod joint
13. Scraper rings	27. Clamp
14. Seal ring	28. Tie rod

Disconnect hydraulic hoses at steering cylinder and cover openings to prevent entrance of dirt. Remove five cap screws (4), then pull on left end of piston rod (1) to remove all internal parts from left side. Remove inner and outer guides (18 and 3) and shim (2). Pull piston rod and piston assembly out of sleeve (20). Remove "O" rings (19 and 21) from outer diameter of cylinder sleeve. Remove the two scraper rings (13) and seal ring (14) from piston (12). Remove snap rings (7), washers (8), split rings (9), washers (10) and piston (12). Remove "O" ring (11) from piston rod. Remove wiper seal (6) and piston rod seal (5) from outer guide (3). Then, remove wiper seal (15), piston rod seal (16) and "O" ring (17) from inner guide (18).

Clean and inspect all parts for excessive wear, scoring or other damage and renew as necessary. Reassemble cylinder by reversing the disassembly procedure, keeping the following points in mind: Use all new "O" rings, scraper rings and seals when reassembling. Smaller outer diameter of cylinder sleeve (20) must face toward inner rod guide (18). Install outer guide (3) with original shim (2) and tighten cap screws (4) to 70 N·m (50 ft.-lbs.) torque.

NOTE: If inner guide, outer guide or cylinder sleeve are renewed, the following adjustment and selection of shims (2) will be necessary.

Remove outer guide (3—Fig. 47 or Fig. 48) and shim (2). Install a 3 mm (0.118 inch) lead wire between end of sleeve (20) and outer guide (3), without shim. Tighten cap screws (4) to 70 N·m (50 ft.-lbs.) of torque. Remove cap screws and outer guide and measure thickness of compressed lead wire. Select a shim of this thickness for installation. Actual axial play of sleeve should be 0-0.1 mm (0-0.004 inch). Shims are available in thicknesses from 0.5 mm (0.020 inch) to 1.8 mm (0.070 inch). Reinstall outer guide and selected shim and tighten screws to 120 N·m (85 ft.-lbs.) of torque for 2955 models, 70 N·m (50 ft.-lbs.) of torque for all other models.

When attaching tie rod joints (26—Fig. 47) to piston rod (1), apply Loctite 242 to threads. Tighten rod ends on early 2750 models (before SN 583 919 L) and all 2955 models to 300 N·m (220 ft.-lbs.) torque. Tighten rod ends to 250 N·m (180 ft.-lbs.) of torque for all other models. On all models, lock with lock

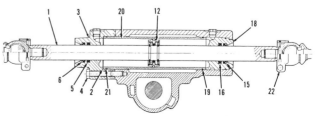

Fig. 48—Cross section of steering cylinder shown in Fig. 47. Cylinder is contained inside casting for axle housing (47—Fig. 14).

plate. Tighten slotted nuts of tie rod ends attached to steering knuckles to 90 N·m (65 ft.-lbs.) torque for M18 slotted nuts; 130 N·m (95 ft.-lbs.) torque for M20 slotted nuts. If necessary, adjust toe-in to 0-3 mm (0-⅛ inch) for all models except 2955, which should be adjusted to 2-5 mm (⁵⁄₆₄-¹³⁄₆₄ inch). Clamp screws to maintain toe-in adjustment should be tightened to 55 N·m (40 ft.-lbs.) torque.

ENGINE AND COMPONENTS

All 2750, 2755 and 2855N models are equipped with 239 cubic inch displacement, four cylinder, turbocharged, direct injection diesel engines.

The 2955 models are equipped with 359 cubic inch displacement, six cylinder, naturally aspirated, direct injection diesel engines.

All these engines have a bore of 106.5 mm (4.19 inches) and a stroke of 110 mm (4.33 inches) and, although they are similar, differences will be noted.

REMOVE AND REINSTALL

All Models

37. To remove engine and clutch assembly, first drain cooling system and, if engine is to be disassembled, drain oil pan. Disconnect batteries and remove front-end weights, if so equipped. Remove side grilles, hood and frame side rails. Disconnect fuel return line and fuel gage wire at fuel tank. Detach wire from air cleaner restriction warning switch and air intake pipe from engine intake manifold. Disconnect hydraulic leak-off line from hydraulic reservoir. Shut off fuel at bottom of tank, then remove fuel line between fuel tank and fuel transfer pump. Disconnect upper and lower radiator hoses and, on models where necessary, remove radiator brace. Relieve pressure, then disconnect hydraulic pump lines, steering lines and oil cooler return line. On models so equipped, detach steering drag link from bellcrank. Loosen hydraulic pump drive shaft clamp screws so front end and pump can be separated from engine. On models equipped with front-wheel drive, unbolt front drive shaft from drive pinion flange. On tractors with air conditioning, loosen condenser attaching screws and pull condenser out from side. On all models, place wooden blocks between front axle and support to prevent tipping. Support tractor under clutch housing and attach a hoist to front end assembly. Remove cap screws that attach front end to engine, then carefully move front end away from engine.

Disconnect speed control rod and shut-off cable at injection pump. Disconnect engine wiring harness, then disconnect cables at starting motor and alternator as required. On models so equipped, detach speed-hour meter cable from clutch housing and disconnect oil pressure warning switch wire at right rear of cylinder block. On all models without Sound Gard Body, unbolt front half of foot rests from flywheel housing. Remove cap screws attaching cowling to flywheel housing. Disconnect starting fluid line at intake manifold. On tractors with Sound Gard Body, disconnect heater hoses from water pump and cylinder block. Remove batteries and battery boxes. Remove platform mat and center section of platform. Through this opening, remove upper left hex nut and upper right cap screw. On tractors with air conditioning, disconnect the two air conditioning lines at couplings. On all tractors, disconnect hydraulic oil reservoir vent line at transmission shift cover. Attach lifting eyes to cylinder head and attach a hoist to lifting eyes. Unbolt flywheel housing from clutch housing and carefully move engine straight forward away from clutch housing.

Reassemble tractor by reversing the disassembly procedure. Recommended tightening torques are as follows:

Clutch housing to engine . . . . 230 N·m
                                              (170 ft.-lbs.)
Drag link to steering
  bellcrank (models so equipped) . . . . 90 N·m
                                              (65 ft.-lbs.)
Front drive shaft to
  drive pinion flange . . . . 75 N·m
                                              (55 ft.-lbs.)
Engine to clutch housing . . . . 230 Nm
                                              (170 ft.-lbs.)

If Torx screws are reused, coat threads with Molykote grease type G before tightening to the recommended torques. The screws attaching the front support to the engine should be tightened carefully as follows:

On 2750 models, tighten the four screws entering the front support from the front and threaded into engine block to 230 N·m (170 ft.-lbs.) torque. Tighten the two screws that enter the engine block flange from the rear and thread into the front support to 180 N·m (130 ft.-lbs.) torque. Install the necessary thickness of shim washers to remove gap between the front support and bosses surrounding the two threaded holes in the oil pan, then tighten the two screws to 400 N·m (300 ft.-lbs.) torque. Refer to the appropriate paragraphs for tightening instructions and recommended torques for other components, such as the hydraulic pump drive.

On 2755 and 2855N models, tighten the six screws attaching the front support to engine block to 230 N•m (170 ft.-lbs.) torque. Install the necessary thickness of shim washers to remove gap between the front support and bosses surrounding the two threaded holes in the oil pan, then tighten the two screws to 400 N•m (300 ft.-lbs.) torque. Tighten the screws securing the side rails to 230 N•m (170 ft.-lbs.) torque. Refer to the appropriate paragraphs for tightening instructions and recommended torques for other components, such as the hydraulic pump drive.

On 2955 models, tighten the four Torx screws attaching the front support to the front of the engine block to 100 N•m (75 ft.-lbs.) torque. The flange of the Torx screws have six markings that correspond to 60 degrees. Mark the front support in line with one flange mark of each of the top two screws, then tighten the top two Torx screws an additional 60 degrees (one mark). Tighten the lower two Torx screws that attach the front support to the engine block to 250 N•m (185 ft.-lbs.) torque. Install the necessary thickness of shim washers to remove gap between the front support and bosses surrounding the two threaded holes in the oil pan, then tighten the two screws to 400 N•m (300 ft.-lbs.) torque. Tighten the eight screws securing the side rails to the front support to 400 N•m (300 ft.-lbs.) torque. Tighten the top two screws (on each side) securing each side rail to the flywheel housing to 325 N•m (240 ft.-lbs.) torque. Tighten the lower screw securing each side rail to the flywheel housing to 575 N•m (420 ft.-lbs.) torque. Refer to the appropriate paragraphs for tightening instructions and recommended torques for other components, such as the hydraulic pump drive.

CYLINDER HEAD

All Models

38. To remove the cylinder head, drain cooling system and remove grille sides and hood. Disconnect battery ground straps. Detach air intake pipe from intake manifold or turbocharger. Remove cross over tube from turbocharger to intake manifold of models so equipped. Detach or remove muffler and exhaust pipe from all models. Unbolt and remove air intake manifold and exhaust manifold. Remove fuel leak-off line, injector lines and fuel injectors. Remove upper coolant hose and thermostat housing. Disconnect fuel lines and remove fuel filter. On tractors equipped with air conditioning, unbolt and remove compressor. Refrigerant lines need not be disconnected. On all tractors, remove ventilator tube and rocker arm cover. Unbolt and remove rocker arm assembly and lift out push rods. Unbolt and remove cylinder head.

NOTE: Make sure cylinder liners are held down with cap screws and washers if crankshaft is to be turned.

Cylinder head should be flat within 0.102 mm (0.004 inch) when measured across head, measured end to end or diagonally. Cylinder head can be surfaced lightly, but thickness should not be less than 104.11 mm (4.099 inches).

Thoroughly clean gasket surfaces of cylinder head and block. Cylinder liner fire ring of early models is 119 mm (4.685 inches) and later model engines have 120 mm (4.724 inches) diameter fire ring. The late cylinder head gasket can be used with early (119 mm) or the later (120 mm) cylinder liners; however, early gasket for smaller (119 mm) fire ring should be used **ONLY** with cylinder liners that have the smaller (119 mm) fire ring. Install new head gasket dry, but oil threads of cylinder head retaining cap screws. Use locating dowels in holes (16 and 17—Fig. 49) of four cylinder models; in holes (24 and 25) of six cylinder models to reduce chance of damage to cylinder head gasket. Make sure hardened flat washers are installed on all cylinder head retaining cap screws without flanged head. Cylinder head retaining cap screws can be reused, but the correct type must be installed and correctly tightened. Use the sequence shown in Fig. 49 and the following identification and tightening data to tighten all cap screws evenly to the correct torque.

Cap screw part number R53223 is 108 mm (4.25 inches) long and is used with a hardened washer in 2750 models. The cylinder blocks that use this screw and washer can be identified by a chamfer or 3 mm (0.118 inch) deep counterbore for each of the screws. Tighten screws in the sequence shown in Fig. 49 in the following four steps.

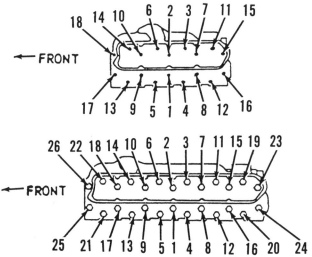

Fig. 49—Tighten cylinder head cap screws evenly in steps using sequence shown. Refer to text.

Step 1 . . . . 47 N•m
(35 ft.-lbs.)
Step 2 . . . . 90 N•m
(65 ft.-lbs.)
Step 3 . . . . 130 N•m
(95 ft.-lbs.)
Step 4 . . . . Break engine
in for 30 minutes at 1500 rpm, then loosen each
cylinder head retaining screw and retorque as in
the previous three steps.

Cap screw part number R85363 is 112 mm (4.41
inches) long, has a flanged head and is used **WITH-
OUT** a hardened washer in 2755, 2855N and 2955
models. Tighten screws in the sequence shown in Fig.
49 in the following four steps.

Step 1 . . . . 100 N•m
(75 ft.-lbs.)
Step 2 . . . . 150 N•m
(110 ft.-lbs.)
Step 3 . . . . Wait 5 minutes,
then recheck to be sure all screws are tightened
to the correct torque listed in Step 2.
Step 4 . . . . Turn each screw
an additional 50-70 degrees. Retightening these
screws after engine break-in is not required.

Tighten rocker arm clamp bolts to a torque of 47
N•m (35 ft.-lbs.). Adjust valve clearance as outlined in
paragraph 40. Install rocker arm cover and gasket
and tighten cover cap screws to 11 N•m (8 ft.-lbs.)
torque. On all models, rocker cover gasket must be
installed dry. **DO NOT** use liquid gasket or adhesive.
Gasket can be reused in most cases if retaining cap
screws are not overtightened. Some models use a
rocker arm cover made of a composite material; additional care should be used when cleaning this material. Acetone can be used to clean groove of composite
cover. Do not overtighten cover retaining screws.

VALVES AND SEATS

All Models

39. The valve seats are integral with the cast iron
cylinder head of some models and seat on renewable
hardened steel inserts of other models. Replacement
seats are available.

Valve face angle is 43½ degrees and seat angle is
45 degrees for all exhaust valves of all models. Face
angle for intake valves of naturally aspirated engines
is 43½ degrees and seat angle is 45 degrees. Face
angle for intake valves of turbocharged models is
29½ degrees and seat angle is 30 degrees. Recommended seat width is 1.5 mm (0.06 inch) for exhaust
valves of all models and intake valves of naturally
aspirated models. Recommended seat width is 2.0

mm (0.08 inch) for the intake valves of turbocharged
models.

Intake and exhaust valve stem diameter is 9.43-
9.46 mm (0.371-0.372 inch) with a recommended
clearance of 0.05-0.10 mm (0.002-0.004 inch) in guide
bores. Valves are available with stem oversizes of
0.08, 0.38 and 0.76 mm (0.003, 0.015, and 0.030 inch)
for use if clearance exceeds 0.15 mm (0.006 inch).

With valves installed in cylinder head, check valve
recession (distance between valve and cylinder head
surface). Valve recession for intake valves should be
0.58-1.19 mm (0.023-0.047 inch). Valve recession for
exhaust valves should be 0.97-1.83 mm (0.038-0.072
inch). If recession exceeds 1.63 mm (0.064 inch) for
intake valves or 2.26 mm (0.089 inch) for exhaust
valves, new valves should be installed and/or seats
must be renewed.

Adjust valve clearance using the procedure outlined in paragraph 40.

40. VALVE CLEARANCE. The two-position
method of valve adjustment is recommended. Turn
engine crankshaft by hand or use a JDE-83 engine
rotation tool until TDC mark is reached. TDC is
determined by using timing pin JDE-81-4 or an 8 mm
(0.320 inch) rod 80 mm (3.15 inches) long inserted
into hole in left side of flywheel housing. The timing
pin will enter hole in flywheel when No. 1 piston is at
TDC. Check the valves of the front No. 1 cylinder to
determine whether it is at top of compression stroke
or at TDC of exhaust stroke. If both valves for the
front cylinder are loose, piston is on compression
stroke. If the valves are tight, rotate crankshaft one
complete revolution and recheck. With No. 1 piston
at TDC of compression stroke, refer to Fig. 50 or Fig.

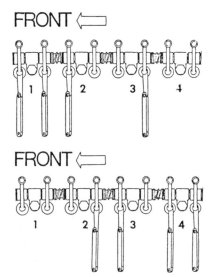

Fig. 50—On four-cylinder models, the valves indicated at
top can be adjusted when No. 1 piston is at TDC of
compression stroke. Turn crankshaft one complete revolution and adjust remaining valves indicated in lower
drawing.

51 and adjust the indicated valves. Turn crankshaft one complete revolution until No. 1 cylinder is at TDC of exhaust stroke, then adjust remaining valves. Valve clearance should be 0.35 mm (0.014 inch) for intake valves and 0.45 mm (0.018 inch) for exhaust valves.

> NOTE: Most engines have a self-locking adjusting screw that should not move during normal engine operation. Some Saran built engines have an Allen head locking set screw located in the rocker arm above the adjusting screw. If a locking screw is used, loosen the lock screw before changing valve clearance. Tighten locking Allen screw to 10 N·m (7 ft.-lbs.) torque after adjusting valve clearance, then recheck valve clearance.

VALVE GUIDES AND SPRINGS

All Models

41. Valve guide bores are an integral part of cylinder head. Valves with 0.08, 0.38 and 0.76 mm (0.003, 0.015 and 0.030 inch) oversize stems are available for service. Guides can also be knurled to reduce clearance with the standard stem. Standard valve guide bore diameter is 9.51-9.53 mm (0.374-0.375 inch) and normal stem clearance is 0.05-0.10 mm (0.002-0.004 inch).

Intake and exhaust valve springs are interchangeable and may be installed either end up. Renew any spring that is distorted, rusted or discolored, or does not conform to the following test specifications:

Free Length . . . . 54 mm
(2.125 in.)

Test at length—
Closed . . . . 240-276 N at 46 mm
(54-62 lbs. at 1.81 in.)
Open . . . . 591-680 N at 34.5 mm
(133-153 lbs. at 1.36 in.)

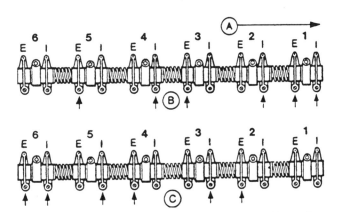

Fig. 51—On six-cylinder models, the valves indicated at top can be adjusted when No. 1 piston is at TDC of compression stroke. Turn crankshaft one complete revolution and adjust remaining valves indicated in lower drawing.

ROCKER ARMS

All Models

42. The rocker arm shaft is retained to the cylinder head by supports attached to the cylinder head by a cap screw through each support. All rocker arms are identical, but should not be relocated on shaft after wear pattern has been established. Recommended clearance between rocker arms and shaft is 0.05-0.13 (0.002-0.005 inch). Bushings in rocker arms are not available separately. Rocker arms must be renewed if bushings are excessively worn.

When reassembling, oil lube hole in rocker arm shaft must be facing downward (toward head) and rearward (toward flywheel). Tighten rocker stand attaching cap screws to a torque of 47 N·m (35 ft.-lbs.). Adjust valve clearance as outlined in paragraph 40.

CAM FOLLOWERS

All Models

43. The barrel-type cam followers can be removed from above after removing cylinder head as outlined in paragraph 38. Cam followers operate in unbushed bores in cylinder block and are available in standard size only.

TIMING GEAR COVER AND FRONT OIL SEAL

All Models

44. The front oil seal contained in timing gear cover can be removed and a new seal can be installed without removing cover from engine. To gain access to the front of the engine is the same as required to remove the timing gear cover. Seal removal and installation is easier with cover removed.

To remove the timing gear cover, first drain cooling system and disconnect battery cables. Remove front end weights, if so equipped. Remove side grilles and hood. Disconnect fuel return line and fuel gage wire at fuel tank. Detach wire from air cleaner restriction warning switch and air intake pipe from engine intake manifold. Disconnect hydraulic leak-off line from hydraulic reservoir. Shut off fuel at bottom of tank, then remove fuel line between fuel tank and fuel transfer pump. Disconnect upper and lower radiator hoses and, on models where necessary, remove radiator brace. Relieve pressure, then disconnect hydraulic pump lines, steering lines and oil cooler return line. On models so equipped, detach steering drag link from bellcrank. Loosen hydraulic pump drive shaft clamp screws so front end can be separated from engine. On models equipped with front-wheel drive, unbolt front drive shaft from drive pinion flange. On tractors with air conditioning, loosen condenser attaching screws and pull condenser out from side. On all models, place wood blocks between front axle and support to prevent tipping. Support tractor under clutch housing and attach a hoist to front end assembly. Remove cap screws that attach front end to engine, then carefully move front end away from engine.

Remove fan, fan belt, alternator and water pump. Remove hydraulic pump drive coupling plate (8—Fig. 52) from crankshaft pulley of models without front pto; coupling (6P or 6PH—Fig. 53) from crankshaft pulley of models with front pto. Remove crankshaft pulley retaining cap screw (9—Fig. 52) from models without front pto, attach a suitable puller and remove pulley (12).

> NOTE: The crankshaft pulley on 6-cylinder models incorporates a vibration damper that should be handled carefully and inspected regularly. The manufacturer suggests that a new damper be installed every 5 years or 4500 hours, whichever occurs first.

Be sure to use a puller that attaches to the inner hub. Never use a jawed puller that pulls against the outer ring. Inspect the outer ring and install a new damper if any movement is detected. A new damper can be painted with a line across the joint between the inner hub, the rubber joint and the outer hub to visually detect any movement of the outer damper ring. Install new damper if outer ring is out of round more than 1.50 mm (0.060 inch). Outer ring wobble should be less than 1.50 mm (0.060 inch) for models without front pto (damper has standard solid mounting to crankshaft). Outer ring wobble should be less than 0.50 mm (0.02 inch) for models with front pto. A collet (12C—Fig. 53) is used to mount the damper to the crankshaft on models with front pto.

To remove pulley from all models with front pto, proceed as follows: Remove pulley retaining screws (12S—Fig. 53) and bump pulley (12P) toward engine to loosen tapered bore from collet (12C). Remove screw (9P) and washer (10), then pull collet (12C) and pulley (12P) from end of crankshaft.

To complete removal of the timing gear cover from all models, remove oil pressure regulating plug, spring and valve. Loosen all oil pan attaching cap screws, then unbolt and remove timing gear cover.

Fig. 52—Exploded view of pump drive and crankshaft used on models without front pto. Note differences for models with standard- and high-volume hydraulic pumps.

1. Hydraulic pump
2. Clamp screw
3. Pump drive shaft
3H. Drive shaft for high-volume pump
4. Cap screws
5. Coupling
6. Cushion
7. Allen screw
8. Adapter coupling
9. Center screw
10. Washer
11. Shaft key
12. Pulley
13. Oil seal
14. Seal sleeve
15. "O" ring
16. Oil slinger
17. Crankshaft timing gear
18. Drive key
19. Crankshaft

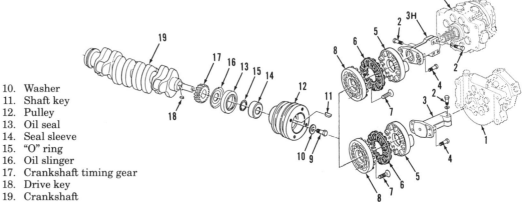

Fig. 53—Exploded view of pump drive and crankshaft used on models with front pto. Note parts (3P and 3S) are different for models with standard- and high-volume hydraulic pumps.

1P. Hydraulic pump with through shaft
1PH. High-volume pump
2A. Set screws
3P. Pump drive shaft
3PH. Drive shaft for high-volume pump
3S. Snap ring
4A. Allen screws
6B. Bushings
6P. Coupling
6PH. Coupling (high-volume pump)
7A. Allen screws
9P. Center screw
10. Washer
12C. Coupling
12P. Pulley
12S. Coupling screws
13. Oil seal
14. Seal sleeve
15. "O" ring
16. Oil slinger
17. Crankshaft timing gear
18. Drive key
19. Crankshaft

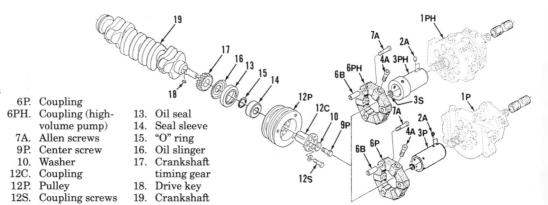

The lip-type front oil seal seats on hub of crankshaft pulley (12—Fig. 52) of early 2750 models or a steel wear sleeve (14—Fig. 52 or Fig. 53) pressed on crankshaft of all later models. Oil slinger (16) is located on crankshaft with cup side toward front as shown. Clean crankshaft, install new "O" ring (15), coat inner surface of new wear sleeve with Loctite 609 or equivalent non-hardening sealer and install on crankshaft.

Open side of front oil seal should be toward inside (rear) of front cover. Coat outer edge of seal with Loctite 609 or equivalent non-hardening sealer and drive seal into bore.

When installing timing gear cover, tighten cover retaining cap screws to 47 N·m (35 ft.-lbs.) of torque. Tighten crankshaft pulley or collet retaining cap screw (9) to 150 N·m (110 ft.-lbs.). Tighten pulley to collet screws (12S—Fig. 53), on models with front pto, evenly to 35 N·m (25 ft.-lbs.) torque. Check alignment of crankshaft pulley after installation, using a dial indicator. Complete the assembly by reversing disassembly procedure. Tightening torques are as follows:

Drag link to steering
bellcrank (models so equipped) . . . . 75 N·m
(55 ft.-lbs.)

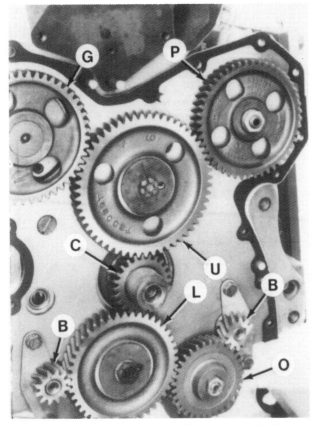

Fig. 54—View of timing gear train assembled. Balance shafts and gears (B) are used only on four-cylinder models.

B. Balance shaft gears
C. Crankshaft gear
G. Camshaft gear
L. Lower idler gear
O. Oil pump gear
P. Injection pump gear
U. Upper idler gear

Front drive shaft to
drive pinion flange . . . . 75 N·m
(55 ft.-lbs.)
Hydraulic pump drive
shaft clamp . . . . 50 N·m
(35 ft.-lbs.)
Oil pan to clutch housing . . . . 47 N·m
(35 ft.-lbs.)
Oil pan to engine block . . . . 47 N·m
(35 ft.-lbs.)
Oil pan to timing gear cover . . . . 35 N·m
(27 ft.-lbs.)

If Torx screws are reused, coat threads with Molykote grease type G before tightening to the recommended torques. The screws attaching the front support to the engine should be tightened carefully as follows:

On 2750 models, tighten the four screws entering the front support from the front and threaded into engine block to 230 N·m (170 ft.-lbs.) torque. Tighten the two screws that enter the engine block flange from the rear and thread into the front support to 180 N·m (130 ft.-lbs.) torque. Install the necessary thickness of shim washers to remove gap between the front support and bosses surrounding the two threaded holes in the oil pan, then tighten the two screws to 400 N·m (300 ft.-lbs.) torque. Refer to the appropriate paragraphs for tightening instructions and recommended torques for other components, such as the hydraulic pump drive.

On 2755 and 2855N models, tighten the six screws attaching the front support to engine block to 230 N·m (170 ft.-lbs.) torque. Install the necessary thickness of shim washers to remove gap between the front support and bosses surrounding the two threaded holes in the oil pan, then tighten the two screws to 400 N·m (300 ft.-lbs.) torque. Tighten the screws securing the side rails to 230 N·m (170 ft.-lbs.) torque. Refer to the appropriate paragraphs for tightening instructions and recommended torques for other components, such as the hydraulic pump drive.

On 2955 models, tighten the four Torx screws attaching the front support to the front of the engine block to 100 N·m (75 ft.-lbs.) torque. The flange of the Torx screws have six markings that correspond to 60 degrees. Mark the front support in line with one flange mark of each of the top two screws, then tighten the top two Torx screws an additional 60 degrees (one mark). Tighten the lower two Torx screws that attach the front support to the engine block to 250 N·m (185 ft.-lbs.) torque. Install the necessary thickness of shim washers to remove gap between the front support and bosses surrounding the two threaded holes in the oil pan, then tighten the two screws to 400 N·m (300 ft.-lbs.) torque. Tighten the eight screws securing the side rails to the front support to 400 N•m (300 ft.-lbs.) torque. Tighten the top two screws (on each side) securing each side rail to the flywheel housing to 325 N•m (240 ft.-lbs.) torque. Tighten the lower screw securing each side rail to the flywheel housing to 575 N•m (420 ft.-lbs.) torque. Refer to the appropriate paragraphs for tightening instructions and recommended torques for other components, such as the hydraulic pump drive.

CAMSHAFT AND TIMING GEARS

All Models

45. The timing gear train (Fig. 54) consists of crankshaft gear, camshaft gear, injection pump gear, oil pump gear, upper idler gear and lower idler gear. Four-cylinder engines also have two balance shaft gears that must be correctly timed to the remaining gears.

The camshaft must be timed with mark on rim of gear closest to the crankshaft and on a line that passes through the center of camshaft and crankshaft. Crankshaft must be positioned so that No. 1 piston is exactly at TDC on compression stroke.

The injection pump gear is marked similarly to the camshaft gear and the mark on the pump gear should also be closest to the crankshaft gear and on a line that passes through the center of injection pump and crankshaft. Crankshaft must be positioned so No. 1 piston is at exactly TDC on compression stroke.

The easiest way to ensure correct timing of all models is to turn crankshaft so that No. 1 piston is at TDC of compression stroke. Remove upper idler gear, align camshaft timing mark, align injection pump timing mark, then reinstall upper idler gear without disturbing crankshaft, camshaft or injection pump gears. Special tool JD-254 or equivalent is available for checking alignment of timing marks.

To time balance shafts of 4-cylinder models, set engine at TDC before installing the lower idler gear and oil pump gears. Timing marks must be in line with center line between crankshaft and balance shaft. Be sure to check to make sure that timing marks are aligned after idler gear and oil pump gear are installed. Refer to Fig. 54.

46. CAMSHAFT AND GEAR. To remove camshaft, first remove timing gear cover as outlined in paragraph 44 and cylinder head as in paragraph 38. Remove oil pan, cam followers and fuel transfer pump.

Before removing camshaft, mount a dial indicator as shown in Fig. 55 and measure camshaft end play. Recommended end play is 0.05-0.23 mm (0.002-0.009 inch) with a wear limit of 0.38 mm (0.015 inch).

Turn camshaft so that thrust plate retaining cap screws are accessible through holes in camshaft gear,

then remove the two retaining cap screws. Pull camshaft and thrust plate from engine.

Camshaft operates in unbushed bores in cylinder block. The camshaft gear is keyed and pressed onto front of shaft. Check camshaft and associated parts against the values that follow:

Camshaft journal diameter—
  Desired . . . . 55.87-55.90 mm
                                             (2.200-2.201 in.)
  Wear limit. . . . . 55.845 mm
                                                     (2.199 in.)
Camshaft bearing bores—
  Diameter. 55.98-56.01 mm
                                             (2.204-2.205 in.)
Camshaft to bearing diametral clearance—
  Desired . . . . 0.010-0.015 mm
                                             (0.004-0.006 in.)
  Wear limit. . . . . 0.18 mm
                                                 (0.007 in.)
Camshaft thrust plate thickness—
  Desired . . . . 3.96-4.01 mm
                                           (0.156-0.158 in.)
  Wear limit. . . . . 3.83 mm
                                                (0.151 in.)
Camshaft **intake** lobe height—
  4 Cyl. engines before SN. CD656457 or TO133255
    Desired . . . . 6.76-7.26 mm
                                        (0.266-0.286 in.)
    Wear limit . . . . 6.50 mm
                                         (0.256 in.)
  4 Cyl. engines after SN. CD656456 or TO133254
    Desired . . . . 6.93-7.42 mm
                                        (0.273-0.292 in.)
    Wear limit . . . . 6.68 mm
                                         (0.263 in.)
  6 Cyl. engines before SN. CD656606 or TO134869
    Desired . . . . 6.76-7.26 mm
                                        (0.266-0.286 in.)

Fig. 55—Check camshaft end play as shown before removing camshaft.

Wear limit . . . . 6.50 mm
(0.256 in.)
6 Cyl. engines after SN. CD656605 or TO134868
Desired . . . . 6.93-7.42 mm
(0.273-0.292 in.)
Wear limit . . . . 6.68 mm
(0.263 in.)
Camshaft **exhaust** lobe height—
All models
Desired . . . . 6.76-7.26 mm
(0.266-0.286 in.)
Wear limit . . . . 6.50 mm
(0.256 in.)

Support camshaft gear and press camshaft from gear if renewal of parts is required. When installing camshaft gear, be sure that thrust plate is in position and that timing mark is toward front. Reinstall camshaft as follows: The easiest method of aligning timing marks is to remove upper idler gear, turn crankshaft until No. 1 piston is at TDC of compression stroke and align camshaft gear and injection pump drive gear timing marks as shown in Fig. 56 and Fig. 57. Then, carefully install upper idler gear without moving crankshaft, camshaft or injection pump drive gears.

Tighten camshaft thrust plate cap screws to a torque of 47 N•m (35 ft.-lbs.). Cap screw for upper idler gear should be tightened to 90 N•m (65 ft.-lbs.) torque.

47. IDLER GEARS. Upper and lower idler gears (U and L—Fig. 54) are bushed and operate on stationary shafts attached to engine front plate with cap screws. Idler gear end play is controlled by thrust washers. Both idler gears are driven by the crankshaft gear (C), and upper idler gear (U) drives camshaft gear (G) and injection pump drive gear (P). Lower idler gear (L) drives oil pump drive gear (O) and right side balance shaft of four-cylinder models.

To remove both idler gears, remove oil pan and timing gear cover. Remove cap screws and pull gears and thrust washers from shafts. Idler gear shafts can now be removed.

Clean and inspect idler gears and shafts and check against the following specifications:

Shaft OD. . . . . 44.43-44.46 mm
(1.749-1.750 in.
Bushing ID . . . . 44.48-44.53 mm
(1.751-1.753 in.)
Operating clearance . . . . 0.02-0.10 mm
(0.001-0.004 in.)
Maximum allowable clearance . . . . 0.15 mm
(0.006 in.)
End play . . . . 0.14-0.29 mm
(0.006-0.012 in.)
Maximum allowable end play . . . . 0.40 mm
(0.015 in.)

Reinstall by reversing removal procedure and be sure camshaft, injection pump and crankshaft are timed as indicated in Fig. 56 and Fig. 57. Tighten cap screw retaining upper idler gear and shaft to 90 N•m (65 ft.-lbs.) torque. If lower idler gear and shaft are retained by a cap screw, tighten to 130 N•m (95 ft.-lbs.) torque; if retained by a hex nut, tighten to 100 N•m (75 ft.-lbs.) torque. Refer to paragraph 50 for correctly timing the balance shafts on four-cylinder models.

48. CRANKSHAFT GEAR. Renewal of crankshaft gear (C—Fig. 54) requires removal of crankshaft as outlined in paragraph 58. Gear is keyed and pressed on crankshaft. Support crankshaft under first throw when installing new gear. Installation of gear may be eased by heating gear in oil to about 182° C (360° F), but extreme care must be exercised to prevent fire, injury or physical damage. Crankshaft gear has no timing marks. Refer to paragraph 46 for timing camshaft and injection pump to crankshaft.

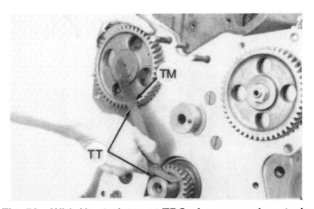

Fig. 56—With No. 1 piston at TDC of compression stroke and timing tool (TT) positioned on shafts centerline as shown, camshaft gear timing mark (TM) should be directly under edge of timing tool.

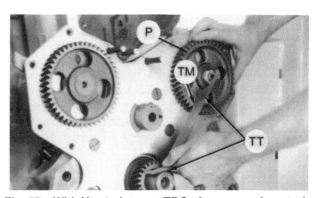

Fig. 57—With No. 1 piston at TDC of compression stroke, injection pump gear (P) is correctly timed if timing mark (TM) is directly below timing tool (TT) when tool is placed between centerlines of crankshaft and injection pump shaft.

Refer to paragraph 50 for correctly timing the balance shaft gears of four-cylinder models.

49. INJECTION PUMP GEAR. To remove the injection pump drive gear, first remove the timing gear cover. The gear is seated on the tapered pump shaft and retained with a hex nut. A Woodruff key in pump shaft ensures correct gear to pump shaft timing. Gear shaft is an integral part of pump and shaft must be removed from gear for pump removal. Tighten gear retaining nut to 80 N·m (60 ft.-lbs.) of torque.

> NOTE: When timing injection pump gear to crankshaft gear, use timing mark indicating number of cylinders of engine. Gear is interchangeable on 3-, 4- or 6-cylinder engines.

50. BALANCE SHAFTS. Four-cylinder engines are equipped with two balance shafts located below the crankshaft on opposite sides of the crankcase. Each shaft is carried in three renewable bushings located in bores in the cylinder block. The right hand balance shaft gear (B—Fig. 54) is driven by the lower idler gear (L) and the left hand balance shaft is driven by the oil pump gear (O). Shafts rotate in opposite directions at twice crankshaft (engine) speed and are designed to dampen the vibration that is inherent in four-cylinder engines.

To remove the balance shafts, first remove the timing gear cover as outlined in paragraph 44. Remove the lower idler gear (L) and oil pump drive gear (O). Identify balance shafts as right and left. On models with bolt-on balance weights, identify the weights for exact placement (left/right, front/rear) on shafts, then unbolt and remove weights. On all models, unbolt thrust plates and carefully withdraw balance shafts from bores.

Use the following specifications to check balance shafts, bushings and thrust plates.

Shaft journal O.D. . . . . 38.137-38.163 mm
(1.5014-1.5024 in.)
Bushing I.D. . . . . 38.177-38.237 mm
(1.5030-1.5054 in.)
Shaft operating clearance—
Desired . . . . 0.02-0.10 mm
(0.001-0.004 in.)
Maximum limit . . . . 0.15 mm
(0.006 in.)
Shaft end play—
Desired . . . . 0.05-0.20 mm
(0.002-0.008 in.)
Thrust plate thickness . . . . 2.97-3.02 mm
(0.117-0.119 in.)

Renew any parts that do not meet specifications. The two front balance shaft bushings can be renewed with engine in tractor; however, the engine, flywheel and flywheel housing must be removed to install rear bushings. Removal of these parts will permit staking the rear bushings.

Use a piloted driver, such as JD-249, to drive front bushings in from front with oil hole aligned with passage in block and with front of bushing flush with chamfer in front of block. Bushings must be staked in position using tool JD-255 or equivalent. Install half round portion of special staking tool in ID of bushing so that the staking ball is in round relief in bushing groove directly opposite oil hole. Locate the square half of special staking tool with correct size dowel engaging the alignment hole in block, then check for correct alignment of cap screw holes in special tool. It may be necessary to turn part of the tool end for end to correctly align the cap screw holes. Install cap screws, check for correct alignment of ball with relief in bushing, dowel with hole and general alignment, then tighten screws evenly until half round part of tool butts against bushing. This procedure will indent the bushing into the dowel hole and prevent it from turning.

If new gears are pressed on shafts, support shaft in holding fixture, position thrust plate over end of shaft, install drive key, then position gear with timing mark out and press gear on shaft. Press gear onto shaft until clearance between thrust plate and gear is within range of desired shaft end play.

Reinstall balance shafts in reverse of removal procedure; however, set engine at TDC before installing the lower idler gear and oil pump gears. Timing marks must be in line with center line between crankshaft and balance shaft. Special tool JD-254 or equivalent is available for checking alignment of timing marks. Be sure that timing marks are aligned after idler gear and oil pump gear are installed.

Tighten thrust plate to block screws to 47 N·m (35 ft.-lbs.) torque and the balance weight attaching nut to 60 N·m (44 ft.-lbs.) torque. If lower idler gear is retained by a cap screw, tighten to 130 N·m (95 ft.-lbs.) torque. If lower idler gear is retained by a nut, secure with thread lock and tighten to 100 N·m (75 ft.-lbs.) torque. Tighten oil pump drive gear to 100 N·m (75 ft.-lbs.) torque. Stake nut in three locations around threads with center punch to stop loosening.

51. TIMING GEAR BACKLASH. Excessive timing gear backlash may be corrected by renewing the gears concerned, or in some instances by renewing idler gear bushing and/or shaft. Refer to the following specifications:

Crankshaft gear to upper idler—
Desired . . . . 0.07-0.30 mm
(0.003-0.012 in.)
Maximum allowable . . . . 0.4 mm
(0.016 in.)

Camshaft gear to upper idler—
Desired . . . . 0.07-0.35 mm
(0.003-0.014 in.)
Maximum allowable. . . . . 0.5 mm
(0.020 in.)
Injection pump gear to upper idler—
Desired . . . . 0.07-0.35 mm
(0.003-0.014 in.)
Maximum allowable. . . . . 0.5 mm
(0.020 in.)
Crankshaft gear to lower idler—
Desired . . . . 0.07-0.35 mm
(0.003-0.014 in.)
Maximum allowable. . . . . 0.5 mm
(0.020 in.)
Oil pump gear to lower idler—
Desired . . . . 0.04-0.36 mm
(0.0016-0.015 in.)
Maximum allowable. . . . . 0.4 mm
(0.016 in.)
4 Cyl. balance gear to lower idler—
Desired . . . . 0.05-0.40 mm
(0.002-0.016 in.)
Maximum allowable. . . . . 0.51 mm
(0.020 in.)
4 Cyl. balance gear to oil pump gear—
Desired . . . . 0.05-0.36 mm
(0.002-0.014 in.)
Maximum allowable. . . . . 0.51 mm
(0.020 in.)

ROD AND PISTON UNITS

All Models

52. Piston and connecting rod assemblies are removed from above after removing cylinder head and oil pan. Secure cylinder liners (sleeves) in cylinder block using short screws and ⅛-inch thick washers to prevent liners from moving as crankshaft is turned. At least five screw and washer combinations should be used on 4-cylinder engines; seven screw and washer combinations should be used on 6-cylinder engines. Sleeve retaining screws should be tightened to 68 N·m (50 ft.-lbs.) torque.

Damage to one piston, piston pin or connecting rod bearing may be caused by the piston cooling orifice that is plugged. The orifices are located in block web near main journals. Tighten orifices to 10 N·m (7 ft.-lbs.) torque.

Pistons are stamped "FRONT" or have an arrow on the piston head or a stamped "B" or "H" on underside of piston pin boss. Connecting rods also have the word "FRONT" stamped (embossed) in the web of connecting rod. **Do not stamp head of unmarked pistons for identification before checking underside of piston boss, because upper ring groove may be damaged.** If necessary to mark piston, use method

that does not result in damage to piston. If mark is not visible on piston, combustion bowl of piston should be offset **away from** camshaft side of engine. Connecting rods and pistons may not be originally numbered, but should be stamped with correct cylinder number before removal.

It is suggested that new connecting rod cap retaining screws be used when making final engine assembly. The previously used screws can be installed to check bearing clearance before final assembly. Two different connecting rod cap retaining screws are used and must be installed in proper connecting rod and tightened to proper torque. Phosphate-coated cap screws, part number R74194, are 54 mm (2.13 inches) long and should be installed in connecting rods marked by two embedded spot facings on rod cap. Cap screws, part number R80033, are 59 mm (2.32 inches) long and should be installed in connecting rod with cap identified by two protruding spot facings.

Both types of connecting rod screws should be dipped in clean oil and tightened evenly to 65-75 N·m (50-55 ft.-lbs.) torque upon final assembly.

PISTONS, RINGS AND SLEEVES

All Models

53. Pistons are cam ground, forged aluminum alloy and are fitted with three rings located above the piston pin. Pistons are available in standard size only, but pistons with different ring height and different head height are used for different applications. Be sure correct type is installed for engine and application. Pistons are stamped "FRONT," marked by an arrow on the piston head or a stamped "B" or "H" on underside of piston pin boss. **Do not stamp head of unmarked pistons for identification before checking underside of piston boss because upper ring groove may be damaged.** If necessary to mark piston, use method that does not result in damage to piston.

The difference in ring groove position can be determined by measuring distance between top of top ring groove and top of piston. Distance on low ring pistons is 18 mm (0.709 inch); 13 mm (0.512 inch) for intermediate ring position pistons; 4 mm (0.158 inch) for pistons with high ring position. Pistons with ring grooves near the top of piston may be safely installed in most models, but all pistons in any engine must be the same. **Be sure to check with manufacturer for correct application.**

Mark "B" on piston indicates standard height from centerline of piston pin to top of piston and "H" mark indicates that head of piston is 0.13 mm (0.0059 inch) higher than standard. The different heights permit control of the compression space, but replacement should normally be with same type that was originally installed.

Piston should be assembled to connecting rod with mark ("FRONT," arrow, "B" or "H") on piston and "FRONT" stamped side of connecting rod toward front of engine. If marks on piston are not visible, combustion bowl of piston should be offset **away from** camshaft side of engine.

Top piston ring is of keystone design and a wear gage (JDE-62) should be used for checking piston groove wear. Normal side clearance for middle (second) piston ring is 0.038-0.076 mm (0.0015-0.003 inch) with a maximum wear limit of 0.2 mm (0.008 inch). Installation instructions for piston rings are included in ring kits.

The renewable wet-type cylinder sleeves are available in standard size only. Sleeve flange at upper edge is sealed by the head gasket. Sleeves are sealed at lower edge by packing and two "O" rings (Fig. 58). Sleeves normally require loosening with a sleeve puller; then they can be withdrawn by hand. Out-of-round or taper should not exceed 0.05 mm (0.002 inch). Check sleeves carefully for rust pits or cracks. If sleeve is to be reused, it should be deglazed, leaving a normal cross-hatch pattern.

When installing sleeves, first make sure sleeve and block bore are absolutely clean and dry. Carefully remove any rust or scale from seating surfaces and from any other areas where loose scale might interfere with sleeve or packing installation. Clean all seating surfaces for packing and "O" rings, and grooves in block for "O" rings. If sleeves are being reused, buff rust and scale from outside of sleeve.

Install sleeve without the seals and measure standout. See Fig. 58. Check sleeve stand-out at several locations around sleeve flange. Sleeve stand-out should be 0.01-0.10 mm (0.0004-0.004 inch) and should be less than 0.06 mm (0.002 inch) difference between adjacent cylinders. If sleeve stand-out is less than minimum specification, install one special shim between sleeve flange and cylinder block. Shims are available in thicknesses of 0.05 and 0.10 mm (0.002 and 0.004 inch). **Do not install more than one shim.** If stand-out is more than 0.10 mm (0.004 inch), check for scale or burrs under flange. Then, if necessary, select another sleeve. After matching sleeves to all the bores, **mark the cylinder number location on the top of fire ring at front of sleeve** to facilitate assembly and later identification. **DO NOT stamp lip of sleeve.** Also record the measured stand-out of each selected sleeve so that this measurement can be compared to stand-out after packing and "O" rings are installed. The difference in stand-out measurements will be the amount that packing is compressed. Refer to the appropriate following paragraphs for packing and sleeve installation. Refer to preceding paragraphs for notes on assembling piston to connecting rods. Piston should not protrude more than 0.33 mm (0.013 inch) higher than face of cylinder block. If protrusion is less than 0.18 mm (0.007 inch)

above face of cylinder block with pistons marked "B," pistons marked "H" can be installed.

54. SLEEVE PACKING AND "O" RINGS. Apply liquid soap to the square-sided packing ring (1—Fig. 58) and install over end of cylinder sleeve. Slide packing ring up against shoulder on sleeve. Make sure that ring is not twisted and that longer sides are parallel with side of sleeve. Apply liquid soap to "O" rings (2) and install the **black "O" rings in lower grooves** of block; then, install the **red "O" rings in top grooves**. Be sure "O" rings are completely seated in grooves so that installing sleeve will not damage "O" rings. Install sleeve carefully into correct cylinder block bores. If sleeve was used previously, be sure that installation is in the same cylinder location. If used sleeve is pitted, it is recommended that sleeve be installed 90 degrees from originally installed position so that pitted surface of sleeve is toward front or rear. Cylinder number previously marked on top of sleeve fire ring should normally be toward front of engine. Work sleeve gently into position by hand until it is necessary to tap sleeve into position using a hardwood block and hammer.

NOTE: Be careful not to damage packing ring or "O" rings.

Check sleeve stand-out (Fig. 58) with packing installed. The difference between this measured stand-out and similar measurement taken earlier for same sleeve in same bore without packing, will be the

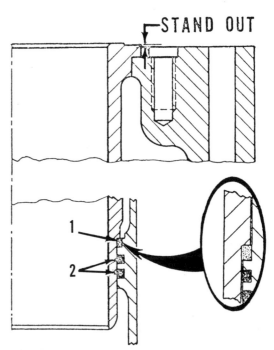

Fig. 58—Sectional view of cylinder sleeve and block showing square-sided packing ring (1) and "O" rings (2). Refer to text for correct installation of cylinder sleeves.

compression of the packing. If compression is less than 0.13 mm (0.005 inch), packing ring will not seal properly. Remove sleeve from cylinder block, check packing ring to be sure installation has not cut the packing ring. If shoulders on sleeve and in cylinder block do not provide proper compression of packing ring, install different sleeve and recheck. If a different sleeve will not provide enough compression of packing ring, suggested repair is to install new cylinder block.

55. SPECIFICATIONS. Specifications of pistons and sleeves are as follows:

Naturally Aspirated 2955 Models
Piston skirt* diameter......... 106.38-106.40 mm
(4.188-4.189 in.)
Sleeve bore diameter.......... 106.48-106.52 mm
(4.192-4.194 in.)
Piston skirt* to sleeve clearance—
Desired .... 0.08-0.14 mm
(0.003-0.005 in.)

All Turbocharged Models
Piston skirt* diameter......... 106.38-106.40 mm
(4.188-4.189 in.)
Sleeve bore diameter.......... 106.48-106.52 mm
(4.192-4.194 in.)
Piston skirt* to sleeve clearance—
Desired .... 0.08-0.15 mm
(0.003-0.006 in.)

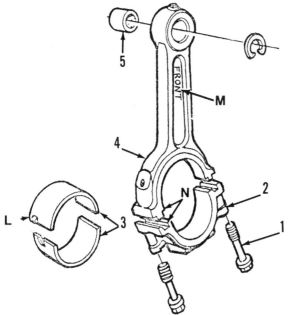

Fig. 59—The diagonal parting line of connecting rod should be away from camshaft side.

1. Cap screws
2. Rod cap
3. Bearing inserts
4. Connecting rod
5. Pin bushing
L. Locating tangs
M. "FRONT" marking
N. Locating notches

*Measured 19 mm (0.740 inch) from bottom of skirt, at right angle to piston pin bore.

PISTON PINS

All Models

56. The full-floating piston pins are retained in pistons by snap rings. Pin bushing (5—Fig. 59) is fitted in upper end of connecting rod and bushing must be reamed after installation to provide a thumb press fit for the piston pin. Piston pin to piston recommended clearance is 0.003-0.023 mm (0.0001-0.0009 inch) for 34.93 mm (1.375 inches) diameter piston pin; 0.005-0.025 mm (0.0002-0.0009 inch) for 41.28 mm (1.625 inches) diameter pin. Wear limit of pin is 34.907 mm (1.3743 inches) or 41.257 mm (1.6243 inches).

Recommended piston pin to bushing clearance is 0.02-0.06 mm (0.0008-0.0024 inch) with a wear limit of 0.10 mm (0.004 inch) for all models.

Damage to one piston, piston pin or connecting rod bearing may be caused by the piston cooling orifice that is plugged. The orifices are located in block web near main journals. Tighten orifices to 10 N•m (7 ft.-lbs.) torque.

CONNECTING RODS AND BEARINGS

All Models

57. The steel-backed, aluminum-lined bearings can be renewed from below after removing oil pan and rod caps. The connecting rod big end parting line is diagonally cut and rod cap is offset away from camshaft side as shown in Fig. 59. A tongue and groove cap joint positively locates the cap. Rod marking "FRONT" should be forward and locating tangs on bearing inserts must be together when cap is installed.

Damage to one piston, piston pin or connecting rod bearing may be caused by the piston cooling orifice that is plugged. The orifices are located in block web near main journals. Tighten orifices to 10 N•m (7 ft.-lbs.) torque.

Connecting rod bearings are available in undersizes of 0.25, 0.51 and 0.76 mm (0.010, 0.020 and 0.030 inch) as well as standard size. Refer to the following specifications:

Crankpin diameter .... 69.80-69.82 mm
(2.748-2.749 in.)
Connecting rod bearing clearance—
Desired .... 0.03-0.10 mm
(0.0012-0.004 in.)
Maximum allowable................. 0.16 mm
(0.0062 in.)

Rod bolts—
Application . . . . Refer to paragraph 52
Torque. . . . . 65-75 N·m
(50-55 ft.-lbs.)

NOTE: Do not reuse connecting rod cap screws. Use new cap screws and dip in clean engine oil when installing. Refer to paragraph 52 for additional notes.

CRANKSHAFT AND BEARINGS

All Models

58. The crankshaft main bearing inserts can be renewed after removing oil pan, oil pump and main bearing caps. All main bearing caps, except rear cap, are alike, but should not be interchanged. Main bearing caps should be numbered from front and must be installed in their original position. Crankshaft end thrust is controlled by three half thrust washers at rear main bearing. Two of the thrust washer halves are located at rear of rear main bearing and one is in front of rear main bearing. Thrust washers are available in standard and 0.17 mm (0.007 inch) oversize. Crankshaft end play should be 0.25-0.33 mm (0.001-0.013 inch) with a maximum end play of 0.50 mm (0.020 inch). Install main bearing caps with identification marks in numerical order from front to rear and the machined pad in arrow shape pointing to camshaft side of engine. Bearing insert locating tabs should be on same side. Tighten main bearing cap retaining screws to 115 N·m (85 ft.-lbs.) of torque.

To remove the crankshaft, it is necessary to remove engine from tractor. Then, remove oil pan, oil pump, cylinder head and rod and piston units. Remove clutch, flywheel and flywheel housing. Remove timing gear cover, camshaft, injection pump drive gear and both idler gears. Then, unbolt and remove engine front plate from cylinder block.

Check crankshaft and bearings for wear, scoring or out-of-round condition using the following specifications:

Main journal
standard diameter . . . . 79.34-79.36 mm
(3.123-3.124 in.)
Journal taper per 25.4 mm (1 in.) length—
Maximum allowable. . . . . 0.0025 mm
(0.0001 in.)
Journal out-of-round—
Maximum allowable. . . . . 0.08 mm
(0.003 in.)
Main bearing diametral clearance—
Desired . . . . 0.03-0.10 mm
(0.0012-0.004 in.)
Maximum allowable. . . . . 0.15 mm
(0.006 in.)

Crankpin diameter . . . . 69.80-69.82 mm
(2.748-2.749 in.)
Connecting rod bearing clearance—
Desired . . . . 0.03-0.10 mm
(0.0012-0.004 in.)
Maximum allowable. . . . . 0.16 mm
(0.0062 in.)

If crankshaft does not meet specifications, either renew it or grind it to the correct undersize. Main bearings are available in standard size and undersizes of 0.25, 0.51 and 0.76 mm (0.010, 0.020 and 0.030 inch). Be sure to check for local availability before machining crankshaft.

Piston cooling orifices, located in block web near main journals, can be removed for cleaning or renewal with crankshaft removed. Each orifice should be clean and open. If orifice is plugged, piston, piston pin or rod bearing cooled by oil jet from the plugged orifice could fail. If orifice is left out upon reassembly, oil pressure will be very low. Tighten orifices to 10 N·m (7 ft.-lbs.) torque.

Refer to paragraph 59 for installation of crankshaft rear oil seal and crankshaft wear ring assembly. **DO NOT attempt to reuse old crankshaft rear oil seal and wear ring.**

CRANKSHAFT REAR OIL SEAL

All Models

59. The lip-type rear oil seal is contained in flywheel housing and a wear ring is pressed on crankshaft flange. The seal and wear ring are pre-assembled and attempts to reinstall seal over wear ring **may result in oil leakage.** A suitable seal installing set, such as JT30040 or KCD1002, should be used to install the seal in flywheel housing and wear ring on crankshaft at the same time.

To renew the seal and wear ring, first remove clutch and flywheel. Seal can be pulled from housing using a jack puller or seal removing tool with tool attached to seal at three equally spaced locations. Wear ring on crankshaft can be removed using a small sharp chisel. Be careful not to damage the crankshaft. Clean crankshaft and inner surface of wear ring with trichloroethylene, then apply a thin coating of Loctite 609 sealer to crankshaft flange. Attach seal installing set pilot to rear of crankshaft, and position seal and wear ring over pilot and on crankshaft. Use seal and pilot installing set to press seal into housing and crankshaft wear ring on to crankshaft simultaneously. Both the open side of the seal and chamfered inner edge of wear ring should be forward, toward inside of engine. **If seal and wear ring are separated, it is suggested that assembly be discarded because improper sealing will probably** result from attempting to reinstall seal over wear ring, even on new seal.

FLYWHEEL

All Models

60. To remove flywheel, first remove clutch as outlined in paragraph 100, then unbolt and remove flywheel from its doweled position on crankshaft. Use two of the flywheel retaining cap screws for jack screws in threaded holes in flywheel and force flywheel off the crankshaft. Use caution to prevent injury or damage when handling the heavy flywheel.

To install a new flywheel ring rear, heat new gear in oil to a temperature of 148° C (300° F). Install ring gear with chamfered end of teeth toward front of flywheel.

Check pilot bearing and renew, if necessary. Install new bearing with shielded side rearward and pack bearing with a high-temperature grease.

When installing flywheel, use new cap screws with Loctite 242 on threads and tighten to 160 N·m (120 ft.-lbs.) of torque. If R74444 flanged head high strength screws are installed, tighten to 55 N·m (40 ft.-lbs.) torque, then turn screws an additional ⅙ turn (60 degrees).

FLYWHEEL HOUSING

All Models

61. The flywheel housing is secured to rear face of engine block by eight cap screws. Flywheel housing contains the crankshaft rear oil seal and oil pressure sending unit switch. The rear camshaft bore in block is open. It is important that gasket between engine block and flywheel housing is in good condition and cap screws properly tightened to prevent leakage. If flywheel housing has "O" ring grooves, use plain

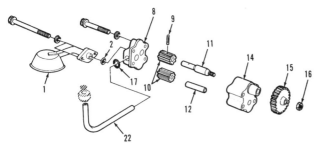

Fig. 60—Exploded view of engine oil pump. Refer to Fig. 63 for oil pressure regulating valve.

1.	Oil pick up	12.	Idler shaft
2.	"O" ring	14.	Housing
8.	Cover	15.	Drive gear
9.	Groove pin	16.	Nut
10.	Pump gears	17.	"O" ring
11.	Drive shaft	22.	Outlet tube

gasket and "O" rings. If housing does not use "O" rings, use gasket with silicone bead. **Do not use adhesives on gasket.** Tighten the ⅜-inch cap screws evenly to 30 N·m (22 ft.-lbs.), then retorque to 47 N·m (35 ft.-lbs.). If so equipped, tighten the four ⅝-inch cap screws to 230 N·m (170 ft.-lbs.) of torque. Refer to paragraph 59 for installation of new crankshaft rear oil seal. Use of old seal or improper installation procedures will probably result in early seal failure.

OIL PUMP AND RELIEF VALVE

All Models

62. OIL PUMP. To remove oil pump, first drain and remove oil pan, then remove timing gear cover as in paragraph 44.

On four-cylinder models, the left balance shaft is driven by the oil pump gear, so the engine should be set at TDC before removing the pump. Turn engine crankshaft by hand or with a JDE-83 engine rotation tool until TDC mark is reached. TDC is determined by using timing pin JDE-81-4 or an 8 mm (0.320 inch) rod 80 mm (3.15 inches) long inserted into hole in left side of flywheel housing. The timing pin will enter hole in flywheel when No. 1 piston is at TDC.

On all models, remove nut retaining oil pump drive gear and using a suitable puller, remove gear. Unbolt and remove oil pump.

With pump removed, use Fig. 60 as a guide and proceed as follows: Remove oil pickup (1) and pump cover (8). Remove idler gear, drive gear and shaft (11) from pump housing. Check to see that groove pin (9) is tight in gear and drive shaft. Pin (9) can be renewed if necessary. Check bearing OD of drive shaft (11), which should be 16.02-16.03 mm (0.630-0.631 inch). Check ID of housing bore for drive shaft (11), which should be 16.05-16.08 mm (0.632-0.633 inch). Renew pump if bore diameter exceeds 16.16 mm (0.636 inch).

Idler gear shaft (12) can be pressed from pump housing if renewal is necessary. Diameter of new shaft is 12.32-12.34 mm (0.485-0.486 inch).

Thickness of new gears is 41.15-41.20 mm (1.62-1.622 inches). Install gears and shafts in pump body as shown in Fig. 61 and measure between ends of gear teeth and pump body. This radial clearance should be 0.10-0.16 mm (0.004-0.006 inch) and pump should be renewed if radial clearance exceeds 0.20 mm (0.008 inch). Place a straightedge across pump body as shown in Fig. 62 and measure distance between straightedge and end of pump gears. This axial clearance should be 0.05-0.17 mm (0.002-0.007 inch) and pump should be renewed if axial clearance exceeds 0.22 mm (0.0085 inch).

Reassemble pump by reversing the disassembly procedure. Install new "O" ring in groove of cylinder block before installing pump. On four-cylinder models, make sure engine is still at TDC, then install oil pump making sure balance shaft drive gear timing mark is aligned with center line from balance shaft to crankshaft after gears are in place. Check keyways in balance shafts, both of which should be at 12 o'clock position.

On all models, tighten screws that attach pump to cylinder block front plate to 47 N·m (35 ft.-lbs.) torque and the pump drive gear retaining nut to 50 N·m (35 ft.-lbs.) torque. Stake nut in three locations around threads with center punch to prevent loosening.

63. OIL PRESSURE RELIEF VALVE. The oil pressure is controlled by a regulating valve located in the forward end of the cylinder block oil gallery. Relief pressure can be adjusted by adding or removing shims (6—Fig. 63) located between spring (5) and plug (7). With engine at 800 rpm and engine lubricating oil at 90° C (194° F), oil pressure should be 100 kPa (14 psi) minimum. The regulating valve spring should have a free length of approximately 120 mm (4.7 inches) and should test 60-75 N (13.5-16.5 lbs.) when compressed to a length of 42.5 mm (1.68 inches). The plug retaining relief valve assembly in timing gear cover should be tightened to 95 N·m (70 ft.-lbs.) torque.

The relief valve seat (bushing) is pressed into the cylinder block as shown in Fig. 64 and is renewable. When installing new valve seat, use JD-248A or equivalent tool to press seat in block until outer recessed edge is flush with bottom of counterbore. Do not press on, or otherwise damage, the raised inner rim of valve seat.

OIL BYPASS VALVE

All Models

64. The oil bypass valve is located behind the engine front plate next to the oil pressure regulating valve. The bypass valve has a set opening pressure and cannot be adjusted. Bypass spring free length should be 59 mm (2.32 inches) and should test 92-112 N (21-25 lbs.) when compressed to a length of 34 mm (1.34 inches).

OIL FILTER

All Models

65. All models are equipped with a full-flow engine oil filter. Manufacturer recommends that oil filter be

Fig. 61—Clearance between gears and housing should be 0.025-0.1 mm (0.001-0.004 inch) when measured as shown.

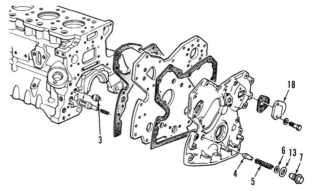

Fig. 63—Exploded view of timing gear cover and front plate showing the engine oil pressure regulating valve.

3. Valve seat (bushing)
4. Regulating valve
5. Spring
6. Shims
7. Plug
13. Aluminum washer
18. Injection pump cover

Fig. 62—Clearance between end of gears and cover should be 0.025-0.15 mm (0.001-0.006 inch) when measured as shown.

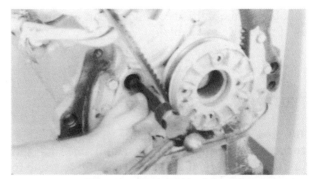

Fig. 64—View showing location of engine oil pressure regulating valve seat. Seat is renewable. Refer to text.

renewed after each 200 hours of operation. Apply a thin film of oil to filter seal ring, then screw filter on until seal ring touches mounting surface. Tighten filter ¾- to 1¼-turns. Do not overtighten.

ENGINE OIL COOLER

All Models

66. The engine oil cooler is mounted externally on right side of engine block and serves also as the oil filter base. Normal servicing of the oil cooler consists of cleaning oil passages and cooler bypass valve with suitable solvent.

TURBOCHARGER

OPERATION

2750, 2755, 2855N Models

67. The exhaust driven turbocharger used on these models supplies air to the intake manifold at above normal atmospheric pressure. The additional air entering the combustion chamber permits an increase in the amount of fuel that can be burned. Burning more fuel will increase the power output of an engine, but it is important that enough air is supplied as well as increasing the amount of fuel.

The use of the engine exhaust to power the compressor increases the engine's flexibility, enabling it to perform with the economy of a smaller engine on light loads yet permitting a substantial horsepower increase at full load. Horsepower loss because of altitude or atmospheric pressure changes is also largely reduced.

The turbocharger contains a rotating shaft with an exhaust turbine wheel on one end and a centrifugal air compressor on the other. The rotating member is precisely balanced and capable of rotative speeds up to 100,000 rpm. The bearings are full-floating sleeve-type and the unit is both lubricated and cooled by a flow of engine oil. Exchange turbocharger units are available, or a qualified technician can overhaul the unit if parts are available.

SERVICE

2750, 2755, 2855N Models

68. In a naturally aspirated diesel engine (without a turbocharger), approximately equal amounts of air enter the cylinders at all loads and only the amount of fuel is varied to compensate for changing power requirements. Turbocharging may supply up to three

times the normal amount of air under full load. All diesel engines operate with an excess of air under light loads. In a naturally aspirated engine, most of the air is used at full load and increasing the amount of fuel results in a higher smoke level with little increase in power output. Excess fuel can also dilute lubrication causing increased wear and damage. Turbocharging can increase the amount of air delivered to the cylinders and the additional air can be used to burn more fuel, resulting in additional power. When more fuel is provided, the turbocharger speed increases and more air is delivered, resulting in more horsepower and heat with little difference in smoke level. Smoke, therefore, cannot be used as a guide to safe maximum fuel setting in a turbocharged engine. **DO NOT** increase fuel volume to increase horsepower above that given in **CONDENSED SERVICE DATA** at the front of this manual.

AiResearch/Garrett or K.K.K. (Kuhnle, Kopp and Kausch) turbochargers are used. Be sure to install correct replacement unit if turbocharger is exchanged. A turbocharger consists of these three main sections: turbine, bearing housing and compressor.

Engine oil taken directly from the clean side of the engine oil filters is circulated through the bearing housing. This oil lubricates the sleeve-type bearings and also acts as a heat barrier between the hot turbine and the compressor. The oil seals used at each end of the shaft are of the piston-ring-type. When servicing the turbocharger, extreme care must be exercised to avoid damaging any of the parts.

> **CAUTION: DO NOT operate the engine, even for a short time, without adequate (abundant) lubrication to the turbocharger unit. When turbocharger is first installed, the engine has not been run for a month or more, or if a new oil filter has been installed, shut fuel OFF, then crank engine with starter until oil pressure is at a safe level. Turn fuel ON after oil pressure reaches normal level, start engine, then operate engine at slow idle speed for at least two minutes before opening throttle or putting a load on engine.**

Additional precautions that should be observed when operating turbocharged engines are as follows:

DO NOT operate at wide-open throttle immediately after starting and slow turbocharger speed down before stopping engine. Allow engine to idle for a while before stopping the engine to slow turbocharger speed. If engine stops suddenly while operating at heavy load and high speed, lack of lubrication to the turbocharger and the high-temperature can damage the turbocharger unit extensively. It is important to restart engine quickly if engine stalls under these conditions.

Maintain air filters and connections. Check condition of the system, especially the restriction indicator, regularly and often. The increased air flow of turbocharged engines may cause filters to become clogged more often than expected. Openings that permit entrance of unfiltered (dirty) air can damage any engine, but damage will occur sooner on turbocharged engines because of the increased volume of air used.

Make sure exhaust pipe opening is closed and the air filter is connected when transporting. If exhaust is equipped with a weathercap, tape the cap closed. If not, tape exhaust pipe closed. If openings are not covered, air passing the open pipe may cause turbocharger to spin without adequate lubrication and damage bearings and other parts.

69. REMOVE AND REINSTALL. To remove the turbocharger, first remove the exhaust pipe, hood and muffler. Disconnect oil lines and air connections from turbocharger. Unbolt turbocharger from the exhaust manifold, then lift unit from tractor.

To inspect the removed turbocharger unit, proceed as follows: Visually inspect turbine wheel and compressor impeller blades for damage by looking through housing and end openings. Refer to INSPECTION AND OVERHAUL paragraphs for specific, measured inspection procedures and specifications.

When installing, attach turbocharger to the exhaust manifold using a new gasket and exhaust adapter ring. Exhaust adapter ring should have 0.8-1.6 mm (0.03-0.06 inch) play. Tighten turbocharger to exhaust manifold attaching screws to 47 N•m (20 ft.-lbs.) torque. Fill the oil drain or inlet port with clean engine oil, then install oil inlet line and oil outlet line to the bearing housing, tightening screws to 27 N•m (20 ft.-lbs.) torque. Reinstall and connect air inlet and pressurized (outlet) pipes. Make sure all parts are properly sealed, tightened and aligned when installing. If necessary to realign the exhaust turbine housing, coat screws with JDT364 NEVER-SEEZ compound or equivalent before installing. Tighten exhaust turbine to bearing housing screws to

7 N•m (5 ft.-lbs.) torque for K.K.K. turbocharger, 14 N•m (10 ft.-lbs.) torque for AiResearch/Garrett turbocharger.

DO NOT operate engine, even for a short time, without adequate (abundant) lubrication to turbocharger unit. When turbocharger is first installed, shut fuel off to prevent starting and crank engine with starter until oil pressure is at a safe level. Turn fuel on after oil pressure reaches normal level, start engine, then operate engine at slow idle speed for at least two minutes before opening throttle or putting a load on engine.

AiResearch/Garrett Turbocharger

70. INSPECTION AND OVERHAUL. Remove the turbocharger unit as outlined in paragraph 69. Before disassembling, mark across compressor housing (1—Fig. 66), bearing housing (13) and turbine housing (17) to aid alignment when reassembling. Clamp turbocharger exhaust inlet flange in a vise and remove cap screws, lockwashers and clamp plates, then remove compressor housing (1). Remove clamp plates, then remove turbine housing (17).

CAUTION: Do not rest weight of any parts on impeller or turbine blades. Weight of only the turbocharger is enough to damage the blades.

To inspect the removed turbocharger unit, proceed as follows: Visually inspect turbine wheel and compressor impeller blades for damage by looking through housing and end openings. Check to see if the compressor or turbine blades have rubbed on the housings. Check to see if housings are wet with oil or have carbon deposits, possibly indicating engine wear or damage. Check turbocharger castings for cracks.

Fig. 66—Exploded view of AiResearch/Garrett turbocharger used on some four-cylinder models.

1. Compressor housing
2. Locknut
3. Compressor impeller
4. Back plate
5. Drain tube
6. Clamp plates
7. Seal ring
8. Thrust collar
9. Thrust bearing
10. Bearing retainers
11. Bearings
12. "O" ring
13. Center housing
14. Lock plate
15. Shroud
16. Turbine shaft & wheel
17. Turbine housing
18. Seal ring
19. Lubrication adapter

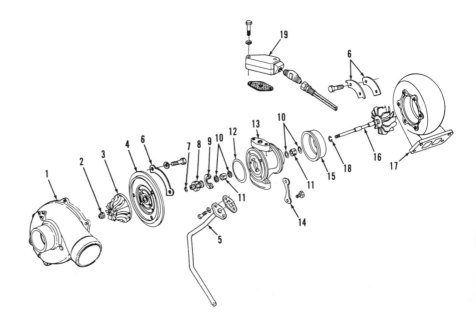

Check in lubricating oil inlet and outlet ports for carbon build up or obvious damage to bearings. Rotate shaft by hand and check for freedom of rotation, without excessive noise or play.

Use a dial indicator with plunger extension to measure radial bearing play through oil outlet port. Move both ends of turbine shaft when checking. Radial play should not exceed 0.08-0.15 mm (0.003-0.006 inch). Move turbine shaft as evenly as possible when checking. Mount dial indicator to end of unit and measure axial (end) play of turbine shaft. Axial play should not exceed 0.25-0.10 mm (0.001-0.004 inch).

> NOTE: Exchange units are available from authorized repair stations and often are an economical alternative to disassembling and repair by inexperienced personnel.

After turbine and compressor housings are removed, hold the turbine shaft from turning using the appropriate holding fixture for turbine wheel (16), then remove locknut (2).

> NOTE: Use a "T" handle or double universal socket to remove lock nut in order to prevent bending turbine shaft.

Lift compressor impeller (3) off, then remove the center housing (13) from turbine shaft while holding shroud (15) onto center housing. Remove back plate (4), thrust bearing (9) and thrust collar (8). Carefully remove bearing retainers (10) from ends and withdraw bearings (11). Spring (Fig. 67) is available as an assembly with back plate (4—Fig. 66).

> CAUTION: Be careful not to damage bearings or surface of center housing when removing retainers. The center two retainers do not have to be

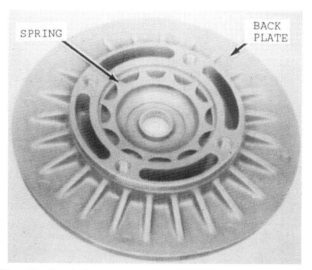

Fig. 67—Backplate and spring must be renewed as an assembly on AiResearch/Garrett models if either part is damaged.

removed unless damaged or unseated. Always renew bearing retainers if removed from grooves in housing.

Clean all parts in a cleaning solution that is not harmful to aluminum. A bristle (not wire) brush and plastic or wood scraper can be used to remove softened deposits. When cleaning, use extreme caution to prevent parts from being nicked, scratched or bent. Parts may be cleaned using a dry bead blast, but pressure should not exceed 280 kPa (40 psi). Make sure all parts are clean, smooth and free of all deposits. Drilled passages should be cleaned with compressed air. Use extreme caution when cleaning to prevent parts from being nicked, scratched or bent.

Inspect bearing bores in center housing (13) for scored surfaces, out-of-round or excessive wear. Standard diameter of bearing bore is 15.819 mm (0.623 inch) and diameter of seal bore is 17.856 mm (0.703 inch). Make certain bore in center housing is not grooved in area where seal (18) rides. Compressor impeller (3) must not show signs of rubbing with either the compressor housing (1) or the back plate (4). Standard diameter of seal bore in back plate (4) is 12.73 mm (0.501 inch). Make certain that impeller blades are not bent, chipped, cracked or eroded. Oil passages in thrust collar (8) must be clean and thrust faces must not be warped or scored. Ring groove shoulders must not have step wear. Standard width of groove in collar (8) for thrust bearing (9) is 4.45 mm (0.175 inch) and standard width of groove for seal ring (7) is 1.67 mm (0.067 inch). Inspect turbine shroud (15) for evidence of turbine wheel rubbing. Turbine wheel (16) should not show evidence of rubbing and vanes must not be bent, cracked, nicked or eroded. Turbine wheel shaft must not show signs of scoring, scratching or overheating. Groove in shaft for seal ring (18) must not be stepped. Standard width of ring groove in shaft (16) for seal ring is 1.867 mm (0.0735 inch). Standard diameter of seal hub of shaft (16) is 17.297 mm (0.681 inch) and diameter of journals is 10.150 mm (0.3994 inch). Check shaft end play and radial clearance when assembling.

Install NEW inner retainers (10), if removed. Lubricate bearings (11) and install in bores of center housing, then install NEW outer retainers (10). **Never reuse the old retainers.** Position the shroud (15) on turbine shaft (16) and install seal ring (18) in groove. Apply a light, even coat of engine oil to shaft journals, compress seal ring (18) with a strong thin tool, such as a dental pick, and install in center housing (13). Install new seal ring (7) in groove of thrust collar (8), then install thrust bearing (9) so that smooth side of bearing is toward seal ring (7) end of collar. Install thrust bearing and collar assembly over shaft, making certain that pins in center housing engage holes in thrust bearing. Install new rubber seal ring (12), make certain that spring (Fig. 67) is positioned in back plate (4—Fig. 66), then install back plate making certain that seal ring (7) is not damaged. Seal ring will be less likely to break if open end of seal ring is installed in bore of back plate first. Install lock plates (14) and screws, tightening screws to 8.5-10.2 N·m (75-90 in.-lbs.) torque. Bend ends of lock plates against heads of screws after all screws are correctly tightened. Install compressor impeller (3) and make certain that impeller is seated against thrust collar (8). Install lock nut (2) and tighten to 2.0-2.3 N·m (18-20 in.-lbs.) torque using a double universal joint socket. Turn the lock nut (2) an additional 90 degrees (or ¼ turn) using a double universal joint socket or "T" handle wrench. The additional 90 degrees should be enough to stretch the turbine shaft 0.140-0.165 mm (0.005-0.0065 inch) to provide the proper tension.

> **CAUTION: Turbine shaft may be bent while tightening nut (2) if "T" handle or double universal joint wrench is not used. Use care even when using the proper tools.**

Install turbine housing (17) with clamp plates (6) next to housing. Coat screws with JDT364 NEVER-SEEZ or equivalent, align previously affixed marks on turbine housing (17) and center housing (13), then tighten screws to 11.3-14.7 N·m (100-130 in.-lbs.) torque. Bend lock plates around screw heads when screws are correctly tightened and turbine housing is properly aligned.

Use a dial indicator to check radial and axial play of turbine shaft at this point of assembly. If axial play is more than 0.102 mm (0.004 inch), thrust collar (8) and/or thrust bearing (9) is worn excessively and unit should be disassembled and new parts installed. Measured end (axial) play of less than 0.025 mm (0.001 inch) indicates incomplete cleaning, dirty assembly or nicked/bent parts and unit should be disassembled and cleaned or repaired before proceeding further. If radial play exceeds 0.15 mm (0.006 inch), shaft (16), bearings (11) or bore in housing (13) is worn excessively and unit should be disassembled and new parts installed.

Install compressor housing (1) with previously affixed marks on housing (1) and center housing (13)

aligned. Clamp screws should be tightened to 7 N·m (5 ft.-lbs.) torque, but final tightening should be accomplished during installation to permit accurate alignment of hose connections.

K.K.K. Turbocharger

71. INSPECTION AND OVERHAUL. Remove the turbocharger unit as outlined in paragraph 69. Before disassembling, mark across compressor housing, bearing housing and turbine housing to aid alignment when reassembling. Clamp turbocharger exhaust inlet flange in a vise and remove cap screws, lock washers and clamp plates, then remove compressor housing. Remove clamp plates, then remove turbine housing.

To inspect the removed turbocharger unit, proceed as follows: Visually inspect turbine wheel and compressor impeller blades for damage by looking through housing and end openings. Check to see if the compressor or turbine blades have rubbed on the housings. Check to see if housings are wet with oil or have carbon deposits, possibly indicating engine wear or damage. Check turbocharger castings for cracks. Check in lubricating oil inlet and outlet ports for carbon build up or obvious damage to bearings. Rotate shaft by hand and check for freedom of rotation, without excessive noise or play. Mount a dial indicator to measure shaft (bearing) radial play at the shaft ends. Radial play should not exceed 0.42 mm (0.016 inch). Move turbine shaft as evenly as possible when checking. Mount dial indicator to end of unit and measure axial (end) play of turbine shaft. Axial play should not exceed 0.16 mm (0.006 inch).

Further disassembly should not be accomplished by other than authorized K.K.K. service stations with necessary parts to replace those found to be worn or damaged. Contact manufacturer or authorized factory service station for parts, repair and information regarding exchange units.

Refer to the service notes in paragraph 69 for installing the turbocharger unit.

FUEL SYSTEM

FUEL LIFT PUMP

All Models

72. The fuel lift pump is mounted on right side of cylinder block and is actuated by a lobe on the engine camshaft. The diaphragm-type pump moves fuel from the fuel tank through the fuel filter to the fuel injec-

tion pump. Some models of pump can be disassembled and cleaned, then reassembled using new diaphragm, screen, spring and gaskets. Parts of pump body, operating levers and check valves are not available separately for any models. No parts are readily available for some pump models and these pumps should not be disassembled. Renewal of complete pump assembly is required if necessary pump repair parts are not available.

WATER TRAP

Models So Equipped

73. A water trap may be installed in the fuel line between the fuel tank and fuel lift pump in left front of radiator. To drain water and dirt from water trap, loosen drain plug (12—Fig. 69), allow sediment to drain, then tighten drain plug finger tight.

To disassemble water trap for cleaning, shut off fuel at tank and remove cap screw (1) with washer (2) and "O" ring (3). Remove base (10) with seal ring (9), glass bowl (8), housing (7), sediment filter (6) and seal rings (5) from body (4).

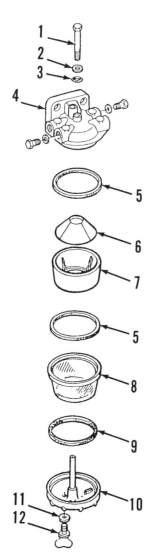

Fig. 69—Exploded view of water trap used on some models.

1. Cap screw	7. Housing
2. Washer	8. Glass bowl
3. "O" ring	9. Seal ring
4. Body	10. Base
5. Seal rings	11. Seal ring
6. Sediment filter	12. Drain plug

Using new seal rings as necessary, reassemble water trap using Fig. 69 as a guide.

FILTER AND BLEEDING

All Models

74. FILTER. All tractors are equipped with a single 2-stage fuel filter. Renew filter element (2—Fig. 70) after every 600 hours of operation. Element should be changed more often if operating under severe conditions or if fuel is contaminated.

To renew the filter element (2), remove drain plug (4), loosen bleed screw (7) and allow filter to drain. Disengage filter element retaining spring (1) and remove element. Make certain filter base (6) is clean, then install new element. Hook top end of retaining spring (1), then hook bottom end. Install drain plug (4). Operate priming lever of fuel lift pump until bubble free fuel flows from bleed screw, then tighten bleed screw (7).

75. BLEEDING SYSTEM. Whenever fuel system has been run dry, fuel filter changed or a line has been disconnected, air must be bled from fuel system as follows: Make certain there is sufficient fuel in tank and that tank outlet valve is open. On models so equipped, loosen drain plug (12—Fig. 69) on water trap and allow any water or sediment to drain. Tighten drain plug finger tight and loosen bleed screw at top of water trap. When fuel flows from bleed screw, tighten bleed screw. On all models, loosen bleed screw (7—Fig. 70) on fuel filter. Operate priming lever of fuel lift pump until bubble free fuel flows from bleed screw, then tighten bleed screw.

Some fuel injection pumps may have bleed screws, but all may be bled as follows: Loosen the fuel supply line to the injection pump, operate the priming lever of the fuel lift pump until bubble free fuel flows from the loosened connection, then tighten the connection.

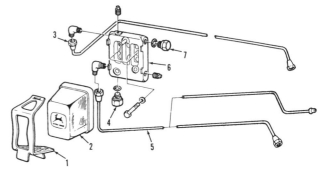

Fig. 70—Exploded view of 2-stage fuel filter typical of all models.

1. Retaining spring	
2. Filter element	5. Inlet line
3. Outlet line	6. Filter base
4. Drain plug	7. Bleed plug

On all models, push priming lever down to lowest position when bleeding the system. If no resistance is felt when operating priming lever of fuel lift pump and no fuel is pumped, lift pump arm is on high point of pump cam on camshaft. In this case, turn engine to reposition pump cam to release pump arm.

If further bleeding is necessary, loosen pressure line connections at injectors one turn, open throttle and crank engine until fuel flows from loosened connections. Tighten connections to 25 N·m (18 ft.-lbs.) torque and start engine.

INJECTOR NOZZLES

All Models

76. LOCATING A FAULTY NOZZLE. If one engine cylinder is misfiring, it is reasonable to suspect a faulty injector. Generally, a faulty injector can be located by running the engine at low idle speed and loosening, one at a time, each high-pressure line at injector. As in checking spark plugs in a spark ignition engine, the faulty unit is the one that least affects the engine operation when its line is loosened.

If a faulty nozzle is found and considerable time has elapsed since injectors have been serviced, it is recommended that all nozzles be removed and checked, or that new or reconditioned units be installed. Refer to the following paragraphs for removal and test procedures.

77. REMOVE AND REINSTALL. To remove an injector, remove hood and wash injector, lines and surrounding area with clean diesel fuel to remove any accumulation of dirt or other foreign material. Dis-

connect leak-off line from injector. Disconnect high-pressure line, then cap all openings. Remove cap screw from nozzle clamp and remove clamp and spacer. Pull injector from cylinder head.

NOTE: Unless the carbon stop seal has failed causing injector to stick, injector can be easily removed by hand. If injector cannot be removed by hand, use special puller JDE-38 and pull injector straight out of bore. DO NOT attempt to pry nozzle from its bore.

Before reinstalling injector nozzle, clean nozzle bore in cylinder head using tool JDE-39, then blow out foreign material with compressed air. Turn tool clockwise only when cleaning nozzle bore. Reverse rotation will dull tool. Install new carbon seal (2—Fig. 71) and seal ring (4) whenever injector has been removed. A protector cap and special pilot (Fig. 72) should be used to push new carbon seal onto nozzle body.

NOTE: Nozzle tip may be cleaned of loose or flaky carbon using a brass wire brush. DO NOT use a brush, scraper or other abrasive on Teflon coated surface of nozzle body between the seals. The Teflon coating may become discolored by use, but discoloration is not harmful.

Insert the dry injector in its bore using a twisting motion. Tighten pressure line connection finger tight, then install hold-down clamp, spacer and cap screw. Tighten cap screw to a torque of 30 N·m (22 ft.-lbs.). Bleed injector if necessary, as outlined in paragraph 75, then tighten pressure line connection to 25 N·m (18 ft.-lbs.) torque. Complete installation by reversing removal procedure.

78. NOZZLE TEST. A complete job of nozzle testing and adjusting requires the use of an approved nozzle tester. Only clean approved testing oil should be used in tester tank. The nozzle should be tested for opening pressure, spray pattern, seat leakage and back leakage. When tested, nozzle should open with

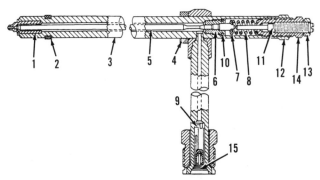

Fig. 71—Sectional view of typical injector assembly. Nozzle tip (1) and valve guide (6) are parts of finished body and are not serviced separately.

1. Nozzle tip
2. Carbon seal
3. Nozzle body
4. Seal ring
5. Nozzle valve
6. Valve guide
7. Spring seat
8. Pressure spring
9. Edge-type filter
10. Seal ring
11. Lift adjusting screw
12. Locknut
13. Locknut
14. Pressure adjusting screw
15. Inlet

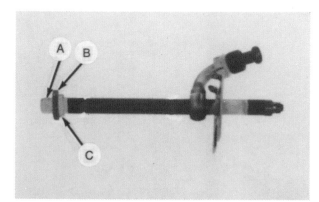

Fig. 72—A protector cap (A) and a new seal washer should be used when installing new carbon seal (C).

a sharp chattering or buzzing sound and cut off quickly at end of injection with a minimum of seat leakage and a controlled amount of back leakage.

Use the tester to check injector as outlined in the following paragraphs:

CAUTION: Fuel leaves the nozzle tip with sufficient force to penetrate the skin. Keep unprotected parts of body clear of nozzle spray when testing.

79. OPENING PRESSURE. Before conducting the test, operate tester lever until fuel flows, then attach injector. Close valve to tester gage and pump tester lever a few quick strokes to be sure nozzle valve is not plugged, that spray holes are open and that possibilities are good that injector can be returned to service without overhaul.

Open valve to tester gage and operate tester lever slowly while observing gage reading. Opening pressure of a used injector should be 20,700 kPa (3000 psi) for naturally aspirated engines; 24,100 kPa (3500 psi) for turbocharged engines. Pressure difference between cylinders should never exceed 700 kPa (100 psi). If pressure is not correct, adjust opening pressure and valve lift as follows:

Loosen lock nut (13—Fig. 71) and while holding pressure adjusting screw (14) from turning, back out lift adjusting screw (11) two turns to ensure against bottoming. Turn pressure adjusting screw (14) until specified opening pressure is obtained. Before tightening lock nut (13), hold pressure adjusting screw from turning, turn lift adjusting screw (11) inward until it bottoms, then back adjusting screw (11) out ½ turn for naturally aspirated engines, ¾ turn for turbocharged engines. Tighten lock nut (13) and recheck opening pressure.

NOTE: When adjusting a new injector or an overhauled injector with a new pressure spring, set opening pressure at 21,700-22,400 kPa (3150-3250 psi) for naturally aspirated engines; 25,100-25,800 kPa (3650-3750 psi) for turbocharged engines. The higher pressure is to allow for initial pressure loss as the spring takes a set.

80. SPRAY PATTERN. The finely atomized nozzle spray should be evenly distributed around the nozzle. Check for clogged or partially clogged orifices or for a wet spray, which indicates a sticking or improperly seating nozzle valve. If spray pattern is not satisfactory, disassemble and overhaul injector as outlined in paragraph 83.

81. SEAT LEAKAGE. Pump tester handle slowly to maintain a gage pressure of 2800-3500 kPa (400-500 psi) below opening pressure while examining nozzle tip for fuel accumulation. If nozzle is in good condition, there should be no wetness for five seconds. If a drop or undue wetness appears on nozzle tip in five seconds, renew injector or overhaul as in paragraph 83.

82. BACK LEAKAGE. Position nozzle on tester so spray tip is slightly higher than adjusting screw end of nozzle. Then, maintain a gage pressure of 10,300 kPa (1500 psi). After first drop falls from adjusting screw, leakage should be at the rate of 3-10 drops in 30 seconds. Leakage of more than 10 drops indicates excessive wear and injector should be renewed. Leakage of less than three drops indicates dirt and varnish build-up and injector should be overhauled.

83. OVERHAUL. First wash the injector in clean diesel fuel and blow dry with compressed air. Remove carbon seal (2—Fig. 71) and seal ring (4). Clean carbon from spray tip using a brass wire brush. Also clean carbon or other deposits from carbon seal groove in injector body. DO NOT use wire brush or other abrasive on the Teflon coating on outside of nozzle body between the seals. Teflon coating can be cleaned with a soft cloth and solvent. Coating may discolor from use, but discoloration is not harmful.

Place nozzle in a holding fixture and clamp the fixture in a vise. NEVER tighten vise jaws on nozzle body without the fixture. Loosen lock nut (13) and back out pressure adjusting screw (14) containing lift adjust screw (11). Remove nozzle body from fixture, invert the body and allow spring (8) and spring seat (7) to fall from nozzle body into your hand. Catch nozzle valve (5) by its stem as it slides from body. If nozzle valve will not slide from body, use special retractor (16481—Fig. 73) to withdraw valve.

Nozzle valve and body are a matched set and should never be intermixed. Keep parts for each injector separate and immerse in clean diesel fuel in a compartmented pan as injector is disassembled.

Clean all parts thoroughly in clean diesel fuel using a brass brush. Hard carbon or varnish can be loosened with a suitable noncorrosive solvent.

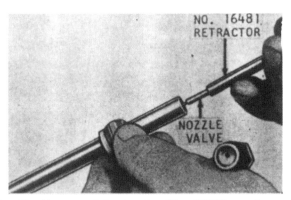

Fig. 73—Use special retractor (No. 16481) as shown to remove sticking nozzle valve.

NOTE: Never use a steel wire brush or emery cloth on spray tip.

On naturally aspirated engines, clean the four 0.28 mm (0.011 inch) spray tip orifices first with a 0.18-0.21 mm (0.007-0.008 inch) cleaning needle held in a pin vise as shown in Fig. 74. Follow up with a 0.25 mm (0.010 inch) cleaning needle.

On turbocharged engines, clean the four 0.30 mm (0.012 inch) spray tip orifices first with a 0.20-0.23 mm (0.008-0.009 inch) cleaning needle held in a pin vise as shown in Fig. 74. Follow up with a 0.27 mm (0.011 inch) cleaning needle.

Clean valve seat using a Valve Tip Scraper and light pressure while rotating scraper. Use a Sac Hole Drill to remove carbon from sac hole.

Piston area of valve can be lightly polished by hand if necessary, using Roosa-Master No. 16489 lapping compound. Use valve retractor to rotate valve. Move valve in and out slightly while turning, but do not apply down pressure while valve tip is in contact with seat.

Valve and seat are ground to a slight interference angle. Seating areas may be cleaned up if necessary using a small amount of 16489 lapping compound, very light pressure and no more than 3 to 5 turns of valve on seat. Thoroughly flush all compound from valve body after polishing.

When reassembling, back lift-adjusting screw (11—Fig. 71) several turns out of pressure-adjusting screw (14), and reverse disassembly procedure using Fig. 71 as a guide. Adjust opening pressure and valve lift as outlined in paragraph 79.

INJECTION PUMP

All Models

84. The Roto-Diesel or CAV/Lucas injection pump is flange mounted on left side of engine front plate and is driven by the upper idler gear of the timing gear train.

Because of the special equipment and specialized training required, service of injection pumps is generally beyond the scope of the average shop. This section will include only the information required for removal, installation and field adjustments of the pump.

85. REMOVE AND REINSTALL. To remove the injection pump, first remove hood, shut off fuel and clean injection pump, lines and surrounding area. Pump can be removed and reinstalled without regard to crankshaft timing position. DO NOT turn crankshaft with injection pump removed.

Disconnect or remove fuel inlet, return and pressure lines, throttle rod and stop cable from injection pump. Drain radiator and remove lower radiator hose. Unbolt and remove access plate from front of timing gear cover. Remove drive gear retaining nut and lock washer. Attach special tool JDG535 or KJD10108 to timing gear cover. Support pump and remove the three mounting stud nuts. Turn center cap screw of special tool until tapered pump shaft is free of drive gear. Remove injection pump, taking care not to lose Woodruff key from pump shaft. Remove special tool.

If the pump shaft does not release easily from the tapered seat of drive gear, it may be necessary to bump the rear face of the engine front cover toward front. Use a large diameter brass drift placed next to the pump mounting flange.

On all models, pump drive gear will be retained by the timing gear cover. DO NOT turn crankshaft with injection pump removed.

To install, turn pump shaft until Woodruff key in shaft aligns with keyway in drive gear. Install lock washer and nut, then tighten nut to a torque of 80 N·m (60 ft.-lbs.). Tighten pump retaining stud nuts finger tight at this time. Rotate top of pump away from cylinder block as far as slotted holes will allow. Then, rotate pump housing toward block until timing marks (3 and 4—Fig. 75) on engine front plate (1) and pump flange (2) are aligned. Tighten pump mounting stud nuts to a torque of 25 N·m (18 ft.-lbs.). Complete

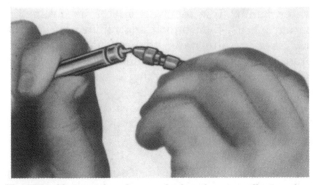

Fig. 74—Use a pin vise and cleaning needle to clean spray tip orifices. Refer to text.

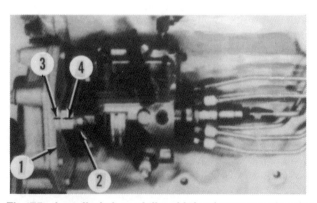

Fig. 75—Installed view of diesel injection pump showing timing marks aligned.

reinstallation by reversing removal procedure. Bleed fuel system as outlined in paragraph 75 and adjust linkage as in paragraph 86.

86. SPEED ADJUSTMENT. Start engine and run until normal operating temperature is reached. Disconnect speed control rod from injection pump and

Fig. 76—Installed view of typical diesel injection pump showing control adjusting points.

R. Speed control rod
1. Low idle screw
2. High-speed screw
3. Pump throttle arm
4. Stop lever

move throttle lever (3—Fig. 76) against high-speed adjusting screw (2) and check engine speed. High-idle speed should be within the range listed in the following table. If necessary, turn high-speed adjusting screw in or out as required. Move throttle lever against low-idle screw (1). If necessary, adjust low-idle screw to obtain recommended engine speed.

High-idle speed—
All 2750 models . . . . 2610-2660 rpm
2755 models w/collar shift trans. . . . 2610-2660 rpm
2755 models w/synchronized trans. . 2410-2510 rpm
All 2855N models. . . . . 2410-2510 rpm
All 2955 models . . . . 2410-2510 rpm

Low-idle speed—
All 2750 models . . . . 700-800 rpm
All Other Models . . . . 750-850 rpm

To adjust shut-off cable, completely push in stop knob and check to be sure stop lever (4) is against stop. Operate engine until warmed up. Pull shut-off knob outward as far as possible and make sure engine stops quickly. Loosen cable clamp and adjust cable if necessary.

COOLING SYSTEM

RADIATOR

All Models

87. REMOVE AND REINSTALL. To remove radiator, first drain cooling system, then remove side grille screens and hood. Disconnect fuel return line, hydraulic leak-off line and wiring harness. Remove air intake pipe. Unbolt fan shroud and lay shroud back over fan. On models with hydraulic oil cooler in front of radiator, unbolt cooler and secure to oil reservoir. On models with hydraulic oil cooler beside the radiator, disconnect and plug hoses to and from cooler, plug openings, then unbolt and remover oil cooler. On models equipped with air conditioning, unbolt condenser and remove from side. On all models, disconnect upper and lower radiator hoses and upper radiator brace, then unbolt and remove radiator.

Reinstall by reversing removal procedure. Capacity of cooling system is as follows:

2750, 2755 & 2855N models—
w/o Sound Gard . . . . 13.0 L
(3.4 gal.)
w/Sound Gard body . . . . 15.0 L
(4.0 gal.)

2955 models—
w/o Sound Gard . . . . 17.0 L
(4.5 gal.)
w/Sound Gard body . . . . 19.0 L
(5.0 gal.)

Radiator cap has a pressure relief opening set at 100-120 kPa (14-17 psi).

WATER PUMP

All Models

88. REMOVE AND REINSTALL. To remove water pump, first remove radiator as outlined in paragraph 87, then unbolt and remove fan and fan belt. Secure alternator to prevent tipping when brackets are detached, disconnect hoses from water pump, then unbolt and remove water pump from engine.

Reinstall by reversing removal procedure. Adjust fan belt so a 90 N (20 lbs.) pull midway between pulleys will deflect belt 19 mm (¾ inch).

89. OVERHAUL. To disassemble water pump, refer to Fig. 77, Fig. 78 or Fig. 79. Unbolt and remove rear cover (2) and gasket (3). Measure distance from rear face of housing (8) to front flanged surface of pulley or hub (10) and record this distance. **Distance changes the alignment of the drive pulleys and misalignment will cause early belt failure and possible damage.** Use a suitable puller to remove pulley or hub (10) from shaft.

If pump impeller (4) has two tapped holes, use the holes and a suitable puller to remove the impeller, then place water pump in a press with impeller side facing upward. Press shaft and bearing assembly (9) out of impeller and pump body. DO NOT attempt to remove shaft and bearing out impeller side of body because housing bore is stepped. Remove ceramic insert (6) and rubber cup (5) from impeller and pull seal (7) from pump body (8).

If pump impeller (4) does not have puller holes, place water pump in a press with impeller side facing down and supported by housing (8). Press shaft, bearing assembly (9), seal (7) and impeller (4) out of pump body (8). If difficulty is encountered, DO NOT attempt to remove shaft and bearing out impeller side of body because housing bore on some models is stepped. Shaft and bearing can be pressed forward out of impeller and housing.

When reassembling, place pump body in a press with front of pump facing upward. A washer with a hole large enough to contact only the outer race of

bearing and not damage the seal should be placed over pump shaft. Press against the washer and outer race of bearing until washer contacts the front of housing. Front of bearing should be flush with front of housing.

Install ceramic insert (6) in rubber cup (5) with "V" groove in insert toward cup. Polished side of ceramic insert should be out toward seal. DO NOT handle ceramic insert with bare hands or seal may leak.

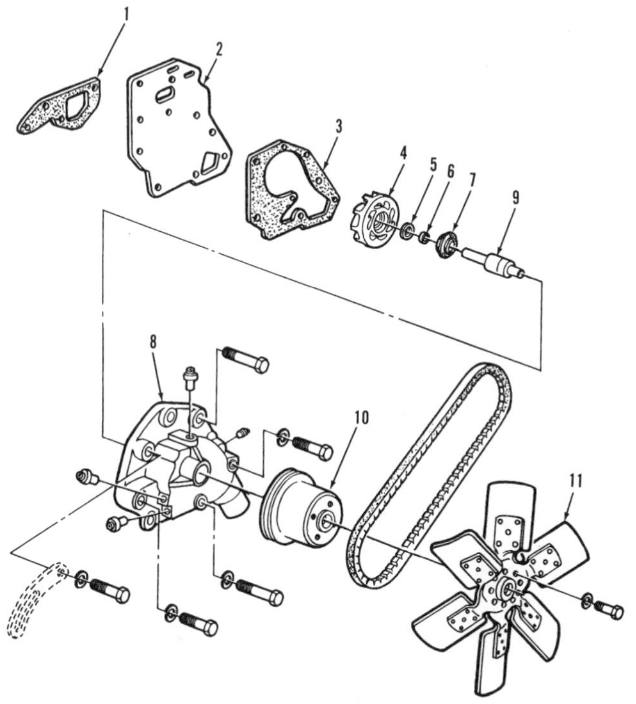

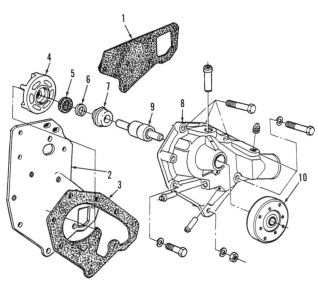

Fig. 78—Exploded view of water pump typical of type used on 2755 and 2855N models. Refer to Fig. 77 for legend.

Fig. 77—Exploded view of water pump typical of type used on 2750 models.

1. Gasket
2. Cover
3. Gasket
4. Impeller
5. Rubber cup
6. Ceramic insert
7. Seal
8. Pump body
9. Shaft & bearing
10. Pulley or hub
11. Fan

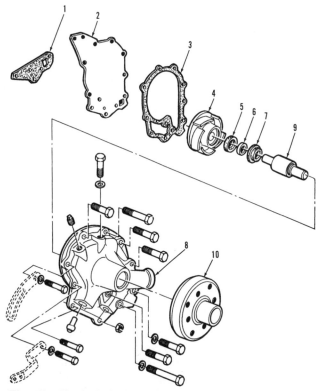

Fig. 79—Exploded view of water pump typical of type used on 2955 models. Refer to Fig. 77 for legend.

Parts must be clean and dry. Place a light coat of engine oil in bore of impeller, then press cup and insert into impeller until cup bottoms in counterbore. Lightly coat outer edge of seal (7) with oil and install in pump body using a socket or other driver that contacts only outer flange of seal. Support front end of pump shaft and press impeller (4) on rear of shaft until rear face of impeller is flush with pump body rear face to 0.25 mm (0.010 inch) below flush.

Invert pump assembly and support the unit on rear of pump shaft in a press. Press pulley or hub (10) on front of pump shaft until front face of pulley or hub is the same distance from rear face of pump body as originally measured before disassembly. Complete the assembly by reversing the disassembly procedure. Suggested distance from rear face of pump housing to the front face of pulley or hub is 162.56 mm (6.40 inches) for all models. Distance changes the alignment of the drive pulleys and misalignment will cause early belt failure and possible damage.

THERMOSTATS

All Models

90. Two thermostats are used on 2955 models; one thermostat is used on other models. Thermostat(s) is located in coolant manifold at front left side of cylinder head. Thermostat(s) should open at 82° C (180° F). It is recommended that thermostat(s) be renewed every two years.

ELECTRICAL SYSTEM

ALTERNATOR AND REGULATOR

All Models

91. OPERATION. All models are equipped with a Bosch 14 volt, 33, 55 or 85 ampere alternator and a regulator that is integral with brush holder.

Model 2750 . . . . 33A R. Bosch 0120339545
55A R. Bosch 0120489704

Model 2755 . . . . 55A R. Bosch 0120489704
55A R. Bosch 0120488218
85A R. Bosch 0120484003
Model 2855N . . . . 55A R. Bosch 0120489704
55A R. Bosch 0120488218
85A R. Bosch 0120484003

Model 2955 . . . . 55A R. Bosch 0120489704
55A R. Bosch 0120488218
85A R. Bosch 0120484003

CAUTION: Because certain components of the alternator can be damaged by procedures that will not affect a DC generator, the following precautions must be observed:

1. When installing batteries or connecting a booster battery, the negative post of battery must be grounded.
2. Never short across any terminal of the alternator or regulator.
3. Do not attempt to polarize the alternator.
4. Disconnect all battery ground straps before removing or installing any electrical unit.
5. Do not operate alternator on an open circuit and be sure all leads are properly connected before starting engine.

Some problems that may occur with the charging system and their possible causes are as follows:

1. Alternator does not charge. Could be caused by:
   a. Slipping fan belt.
   b. Open in charging circuit.
   c. Worn or defective brushes.
   d. Faulty regulator.
   e. Open in rotor field windings.

2. Low or irregular charging output. Could be caused by:
   a. Slipping fan belt.
   b. Intermittent open in charging circuit.
   c. Worn or defective brushes.

3. Excessive charging system voltage. Could be caused by:
   a. Loose alternator connections.
   b. Faulty regulator.

4. Noisy alternator. Could be caused by:
   a. Worn fan belt or misaligned pulley.
   b. Loose pulley.
   c. Worn bearings.
   d. Faulty rectifier diodes.

92. OVERHAUL. To disassemble the alternator, remove brush holder and regulator (12—Fig. 80) and capacitor (13). Immobilize pulley and remove shaft nut, pulley and fan. Mark drive end frame, stator frame and brush end frame for aid in correct reassembly and remove through-bolts. Rotor will remain with drive end frame and stator will remain with brush end frame when alternator is disassembled.

Remove the two terminal nuts and three screws securing rectifier (7) to brush end frame and lift out rectifier and stator as a unit. Identify and tag the three stator leads for correct reassembly, then unsolder leads from rectifier diodes using an electric soldering iron and minimum heat. Be careful not to get solder on diode plates or overheat the diodes.

Check brush contact surface of slip rings for burning, scoring or varnish coating. Surfaces must be true to within 0.03 mm (0.0012 inch) and should have diameter of more than 26.8 mm (1.055 inches). Contact surface can be trued by chucking in a lathe. Polish the contact surface after truing using 400 grit polishing cloth until scratches and machine marks are removed. Maximum allowable radial runout of rotor is 0.05 mm (0.002 inch).

Check continuity of rotor windings using an ohmmeter as shown in Fig. 81 and Fig. 82. Ohmmeter reading should be 4.0-4.4 between the two slip rings and infinity between either slip ring and rotor pole or shaft.

Stator is "Y" wound, the three individual windings being joined in the middle. Test the windings using an ohmmeter as shown in Fig. 83. Ohmmeter reading should be 0.4-0.44 between any two leads and infinity between any lead and stator frame.

Alternator brushes and shaft bearings are designed for 2000 hours service life. New brushes protrude 10 mm (0.4 inch) beyond brush holder when unit is

removed. For maximum service reliability, renew both the brushes and shaft bearings when brushes are worn to within 5 mm (0.2 inch) of holder. Solder copper leads to allow for 10 mm (0.4 inch) protrusion using rosin core solder only. Be sure the solder does not seep into and stiffen the wire lead.

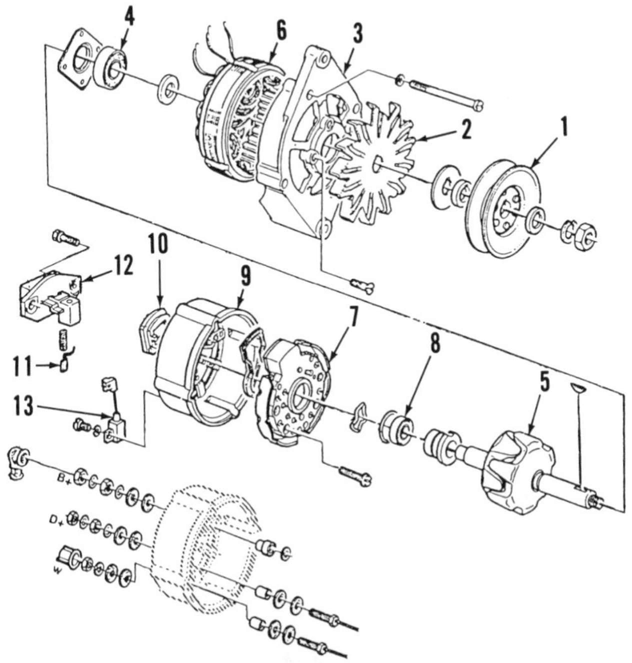

Fig. 81—A reading of 4.0-4.4 ohms should exist between slip rings (1 and 2) when checked with an ohmmeter.

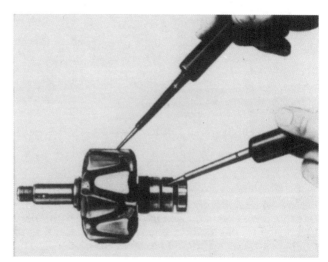

Fig. 82—No continuity should exist between either slip ring and any part of rotor frame.

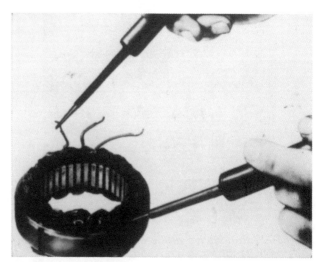

Fig. 83—No continuity should exist between any stator lead and stator frame.

Fig. 80—Exploded view of Bosch 55 ampere alternator. Other Bosch alternators are similar.

1. Pulley
2. Fan
3. Drive end frame
4. Front bearing
5. Rotor
6. Stator
7. Rectifier
8. Rear bearing
9. Brush end frame
10. Insulator
11. Brushes
12. Brush holder & regulator
13. Capacitor

The rectifier is furnished as a complete assembly and diodes are not serviced separately. Rectifier unit contains three positive diodes, three negative diodes and three exciter diodes, which energize rotor coils before engine is started. If any of the diodes fail, rectifier must be renewed.

To test positive diodes, touch positive ohmmeter probe to positive heat sink as shown in Fig. 84, and touch negative test probe to each diode lead in turn. Ohmmeter should read at or near infinity for each test. Reverse the leads and repeat the series. Ohmmeter should read at or near zero for the series.

Test negative diodes as shown in Fig. 85. Place negative test probe on negative heat sink and touch

each diode lead in turn with positive test probe. Ohmmeter should read at or near infinity for the series. Reverse test leads and repeat the test. Ohmmeter should read at or near zero for the series.

Test exciter diodes by using the D+ terminal as the base as shown in Fig. 86. Ohmmeter should read at or near infinity with positive test probe on terminal screw and at or near zero with negative test probe touching screw.

Reassemble alternator by reversing the disassembly procedure. Tighten through-bolts to a torque of 4-5.5 N·m (33-48 in.-lbs.) and pulley nut to 35-65 N·m (25-32 ft.-lbs.) torque.

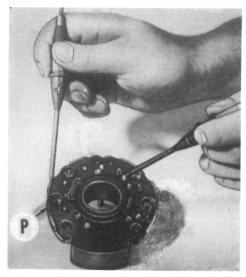

Fig. 84—A near infinity reading should be obtained when positive probe rests on positive heat sink (P) and negative probe touches diode leads as shown. Reverse the probes and reading should be near zero ohms.

STARTING MOTOR

All Models

93. Bosch starting motors are used. Centers in ends of armature shaft may be slightly off-center. Support armature (12—Fig. 87) by outer diameter of shaft ends when checking commutator and armature plate run-out. Original application and test specifications are as follows:

Original Application

Model 2750	R. Bosch 0001359090
	R. Bosch 0001362312
Model 2755	R. Bosch 0001362312
	R. Bosch 0001362316
Model 2855N	R. Bosch 0001362312
	R. Bosch 0001362316
Model 2955	R. Bosch 0001369001
	R. Bosch 0001369005

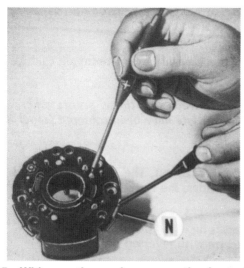

Fig. 85—With negative probe on negative heat sink (N) and positive probe touching diode leads, an infinity reading should be obtained. Reverse the probes and reading should be near zero ohms.

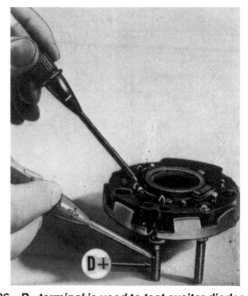

Fig. 86—D+ terminal is used to test exciter diodes. Refer to text.

Robert Bosch Test Specifications

R. Bosch 0001359090
Commutator—
Minimum diameter . . . . 39.5 mm
(1.555 inches)
Maximum out-of-round . . . . 0.03 mm
(0.0012 inch)
Insulation undercut . . . . 0.5-0.8 mm
(0.02-0.03 inch

Armature—
Plate out-of-round, maximum . . . . 0.05 mm
(0.002 inch)
End play . . . . 0.1-0.3 mm
(0.004-0.012 inch)
Brush length minimum . . . . 16 mm
(0.63 inch)

R. Bosch 0001362312, 0001362316, 0001369001
& 0001369005
Commutators—
Minimum diameter . . . . 42.5 mm
(1.67 inches)
Maximum out-of-round . . . . 0.03 mm
(0.0012 inch)
Insulation undercut . . . . 0.5-0.8 mm
(0.02-0.03 inch

Armature—
Plate out-of-round, maximum.. . . . . . . 0.05 mm
(0.002 inch)
End play . . . . 0.1-0.3 mm
(0.003-0.012 inch)

Brush length minimum . . . . 7.5 mm
(0.30 inch)

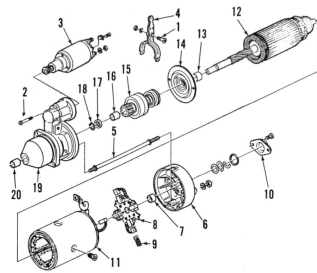

Fig. 87—Exploded view of typical Bosch starter assembly.

1. Screw
2. Screw
3. Solenoid
4. Lever
5. Studs
6. Commutator end housing
7. Bushing
8. Brush holder
9. Springs
10. End cover
11. Field windings assy.
12. Armature
13. Bushing
14. Quill
15. Starter pinion & clutch
16. Bushing
17. Retainer
18. Snap ring
19. Starter nose
20. Bushing

ENGINE CLUTCH

ADJUSTMENT

Models with Mechanical Linkage

94. PEDAL FREE TRAVEL. Clutch pedal free travel (A—Fig. 88) should be 25 mm (1 inch) and should be readjusted when free travel decreases to 13 mm (½ inch). Adjustment is made by changing the length of the clutch operating rod. Loosen locknut, disconnect one end of rod, then thread rod (R) in or out of yoke (Y) as required. Reconnect rod, tighten locknut, then recheck free travel. If clutch disc is new, adjust free travel (A) to 35 mm (1⅜ inches).

All Models with Hydraulic Linkage

95. PEDAL ADJUSTMENT. Remove dash cover attaching screws and lift off left cover. Refer to Fig. 91 and measure distance (B—Fig. 89) between floor plate and pedal pivot point, then measure distance (C) between floor plate and return spring bracket pivot point. Distance (A) is difference between measured distance (B) and measured distance (C). Dis-

tance (A) should be 24-26 mm (0.945-1.023 inches) for all 2750 models, 2755 models before SN 620 388L, 2855N models before SN 620 388L and 2955 models before SN 617 793L. Distance (A) should be 20-22 mm (0.787-0.866 inch) for 2755 models after SN 620 387L, 2855N models after SN 620 387L and 2955 models after SN 617 792L. If adjustment is required, loosen lock nut and turn stop screw (2), which will change

Fig. 88—External clutch adjustment is accomplished by threading yoke (Y) onto or out of rod (R). Refer to text.

measurement (C). Lock adjustment by tightening locknut when adjustment is correct.

To adjust master cylinder operating range on 2750 models, proceed as follows: Move boot (7—Fig. 95) to expose flats on operating rod (8). Loosen lock nut, turn rod in or out of clevis until piston (3) is touching washer (5) and rod (8) is still just contacting piston. Tighten rod lock nut against clevis. Fully depress clutch pedal and measure the distance operating rod (8) travels. Master cylinder operating rod travel should be 26-28 mm (1.02-1.10 inches). Stop screw (A—Fig. 90) can be adjusted to limit master cylinder travel.

To adjust master cylinder operating range for 2755, 2855N and 2955 models, proceed as follows: Move

boot (7—Fig. 96) to expose flats on operating rod (8), then loosen lock nut. Attach a long, clear plastic hose to the bleed screw (Fig. 99), open bleed screw and pump brake fluid into clear plastic hose until level in hose is about 1 meter (40 inches) above the level in reservoir. Make sure that reservoir still has fluid with fluid in hose. Return brake pedal to the original position and turn operating rod (8—Fig. 96) into master cylinder until fluid in hose stops sinking. Turn operating rod out from master cylinder until fluid in hose just begins to sink, then turn rod an additional ½ turn and lock adjustment with lock nut. Adjust stop screw (A—Fig. 90) as follows: Fully depress clutch pedal and measure the distance operating rod (8) travels. Master cylinder operating rod travel should be 26-28 mm (1.02-1.10 inches) for 2755 models before SN 620 388L, 2855N models before SN 620 388L and 2955 models before SN 617 793L. Master cylinder operating rod travel should be 28-30 mm (1.10-1.18 inches) for 2755 models after SN 620 387L, 2855N models after SN 620 387L and 2955 models after SN 617 792L. Position of stop screw (A—Fig. 90) can be adjusted to limit master cylinder travel of all models.

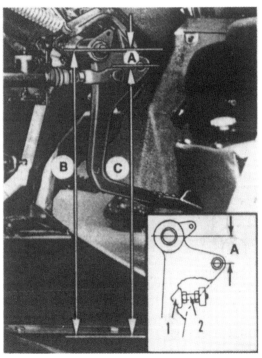

Fig. 89—View showing pedal adjustment typical of models with hydraulically operated clutch. Inset shows measurement (A), pedal stop (1) and stop bolt (2).

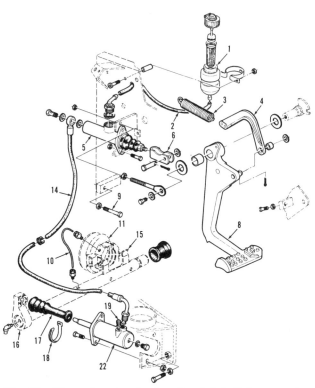

Fig. 91—View showing hydraulic clutch linkage and related parts used on 2750 models with Sound Gard Body.

Fig. 90—Stop screw (A) limits the operating range of master cylinder piston.

1. Reservoir	
2. Hose	11. Release bearing
3. Pedal return spring	14. Pressure hose
4. Return bracket	15. Fork
5. Master cylinder	16. Clutch release shaft
6. Clevis	17. Rubber boot
8. Clutch pedal	18. Tie band
9. Stop bolt	19. Bleeder valve
10. Lube line	22. Slave cylinder

64

With master cylinder and pedal adjusted, external slave cylinder travel can be checked on 2750 models. Fully depress clutch pedal and measure the distance operating rod (12—Fig. 97) travels. This distance should be at least 8.5 mm (5/16 inch) and not more than 12 mm (15/32 inch). If distance is less than 8.5 mm (5/16 inch), bleed system as in paragraph 96 and recheck.

BLEEDING HYDRAULIC CLUTCH SYSTEM

All Models So Equipped

96. To bleed air from system, first fill reservoir with DOT 4 brake fluid. DO NOT use any type of transmission or hydraulic oil. Remove cap from bleed valve (19—Fig. 91 or Fig. 92) and attach a hose to valve. Insert hose end into a container partially filled with brake fluid. Open bleed valve 1/2 turn, fully depress clutch pedal, close bleed valve, then release pedal. Repeat this procedure until only air free fluid flows

Fig. 92—Exploded view of hydraulic clutch linkage and related parts used on 2755, 2855N and 2955 models with Sound Gard Body. Refer to Fig. 91 for legend except the following.

5A. Master cylinder & reservoir
22A. Throwout bearing & slave cylinder

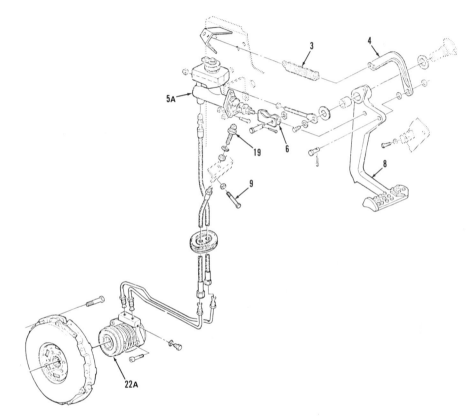

Fig. 93—Slave cylinder is installed at rear of mounting bracket on early 2750 tractors with early style clutch release shaft and release bearing. With new clutch disc installed, secure boot to rod with tie band (1), 9 mm (0.354 inch) between end of boot and front face of cylinder.

Fig. 94—Slave cylinder is installed in front of mounting bracket of late 2750 tractors and earlier tractors if the late style clutch release shaft and release bearing has been installed. With new clutch disc installed, secure boot to rod with tie band (1), 15 mm (0.590 inch) between end of boot and front face of cylinder.

from bleed valve as indicated by flow that is clean and free from bubbles.

If tractor is equipped with Hi-Lo shift unit, remove floor mat and floor panel. Bleed Hi-Lo locking device on transmission shift cover in similar manner.

NOTE: When bleeding system, make sure reservoir does not run dry.

After bleeding system, make sure fluid reservoir is filled.

CLUTCH OPERATING CYLINDERS

2750 Models So Equipped

97. R&R AND OVERHAUL. Remove dash cover attaching screws and lift off left cover. Attach a hose to bleed screw on slave cylinder and insert hose end

in a suitable container. Open bleed screw one turn and drain system by operating clutch pedal. Disconnect lines from cylinders, then unbolt and remove cylinders.

Procedure for disassembly of cylinders is obvious after examination of units and reference to appropriate Fig. 95 or Fig. 97. Repair kits are available for overhauling cylinders.

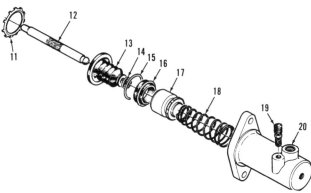

Fig. 97—Exploded view of hydraulic clutch slave cylinder used on 2750 models with Sound Gard Body.

11. Retainer	16. Seal ring
12. Operating rod	17. Piston
13. Rubber boot	18. Spring
14. Retaining ring	19. Bleed valve
15. Snap ring	20. Housing

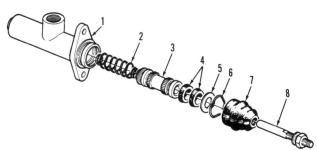

Fig. 95—Exploded view of hydraulic clutch master cylinder used on 2750 models with Sound Gard Body. Refer to Fig. 91 for reservoir (1) and other related parts.

1. Housing	5. Washer
2. Spring	6. Snap ring
3. Piston	7. Rubber boot
4. Seal rings	8. Operating rod

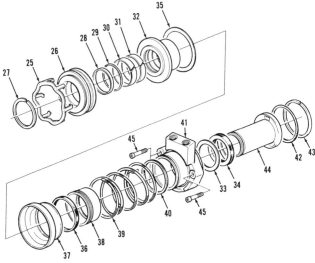

Fig. 98—Exploded view of hydraulic clutch slave cylinder and throwout bearing used on 2755, 2855N and 2955 models with Sound Gard Body. Refer to Fig. 92 for related parts.

25. Retaining ring	36. Seal ring
26. Throwout bearing	37. Outer protective cover
27. Snap ring	38. Inner protective cover
28. Wiper seal	39. Spring
29. "O" ring	40. Seal ring
30. Guide ring	41. Housing
31. Guide ring	42. "O" ring
32. Piston	43. Snap ring
33. Back-up washer	44. Guide sleeve
34. "V" packing	45. Allen screws

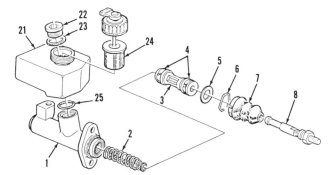

Fig. 96—Exploded view of hydraulic clutch master cylinder used on 2755, 2855N and 2955 models with Sound Gard Body.

1. Housing	
2. Spring	8. Operating rod
3. Piston	21. Reservoir
4. Seal rings	22. Screw
5. Washer	23. Washer
6. Snap ring	24. Strainer
7. Rubber boot	25. "O" ring

Reassemble and reinstall cylinders, check adjustments as in paragraph 95, then fill and bleed system as outlined in paragraph 96.

NOTE: Refer to Fig. 93 and Fig. 94 for installation information of slave cylinders.

2755, 2855N and 2955 Models So Equipped

98. R&R AND OVERHAUL. To remove the clutch master cylinder, remove dash cover attaching screws and lift off left cover. Attach a hose to bleed screw (19—Fig. 92) and insert hose end in a suitable container. Open bleed screw one turn and drain system by operating clutch pedal. Disconnect line from master cylinder, then unbolt and remove cylinder.

Procedure for disassembly of master cylinder is obvious after examination of units and reference to Fig. 96. **DO NOT lubricate master cylinder with anything except DOT 4 brake fluid.** Repair kit is available. Reassemble and reinstall cylinder, check adjustments as in paragraph 95, then fill and bleed system as outlined in paragraph 96.

To remove the slave cylinder, it is necessary to split the tractor as outlined in paragraph 99. Disconnect lines, remove screws (45—Fig. 98), then remove the assembly. To disassemble, remove retaining ring (25) and release bearing (26). Remove snap ring (27), then slide piston (32) from guide sleeve (44). **DO NOT lubricate any of the clutch slave cylinder with anything except DOT 4 brake fluid.** Seals will be damaged by contact with oil or grease. Repair kit is available. Reassemble and reinstall cylinder, rejoin tractor halves, then check adjustments as in paragraph 95. Fill and bleed system as outlined in paragraph 96.

TRACTOR SPLIT

All Models

99. To split engine from clutch housing, drain cooling system. Disconnect and remove batteries. Remove exhaust pipe, radiator side plates and grille screens. On models with Sound Gard Body, remove battery boxes and, on models so equipped, disconnect air conditioning lines at unions under body. On all models, disconnect wires to lights, then remove hood. Disconnect wiring harness from dash panel and engine speed control rod and shut-off cable from injection pump. On models with front-wheel drive, refer to paragraph 7 and remove the front drive shaft. Disconnect heater hoses, hydraulic lift system lines and power steering lines that would interfere with separation of tractor between clutch housing and engine. Detach drag link from steering bellcrank from models so equipped. On models without Sound Gard Body,

remove cap screws that attach dash panel at flywheel housing. On models with an auxiliary fuel tank, drain fuel from main tank, disconnect interfering hoses and remove the auxiliary fuel tank. On all models so equipped, remove the two lower screws that attach the flywheel housing to the oil pan. On all models, attach splitting stands and support both the front and rear sections of tractor in such a way to safely allow separating between clutch housing and flywheel housing. Remove platform mat and center floor panel from models with Sound Gard Body, then remove upper right cap screw through this opening. On all models, remove the cap screws and hex nuts attaching clutch housing to the flywheel housing, then separate between the two housings. **It may be necessary to use special wrench (KJD10129 or equivalent) to remove the upper hex nut from the right side on tractors with Sound Gard Body.**

Refer to appropriate paragraphs for service information of components contained in the housing.

When rejoining tractor, it may be necessary to rotate engine crankshaft (using special tool JDE-83 or equivalent) to facilitate entry of input shafts into clutch discs. Be sure flywheel housing and clutch housing are butted together before tightening retaining cap screws and hex nuts. Recommended tightening torques are as follows:

Clutch housing to engine. . . . . 230 N·m
(170 ft.-lbs.)

Fig. 99—Refer to paragraph 95 to adjust the clutch master cylinder operating rod.

Drag link to steering
   bellcrank (models so equipped) . . . .   90 N·m
                                                    (65 ft.-lbs.)
Front drive shaft to
   drive pinion flange . . . .   75 N·m
                                                    (55 ft.-lbs.)
Engine to clutch housing . . . .   230 N·m
                                                    (170 ft.-lbs.)

Refer to the appropriate paragraphs for tightening instructions and recommended torques for other components, such as the hydraulic pump drive.

R&R AND OVERHAUL CLUTCH

All Models

100. Tractors are equipped with single dry disc diaphragm spring clutch assembly. The 320 mm (12½ inches) diameter LUK clutch is shown in Fig. 100. Some differences may be noticed and it is important to correctly identify the assembly for ordering parts. To remove clutch assembly, split tractor as outlined in paragraph 99, remove hex nuts and cap screws, then lift off clutch assembly.

If flywheel is surfaced, distance from the clutch friction surface of flywheel to clutch cover attaching surface should be 21.35-21.85 mm (0.841-0.860 inch) for 2750 models, 21.47-21.73 mm (0.845-0.855 inch) for 2755, 2855N and 2955 models. The thickness of pressure plate (2—Fig. 100) from diaphragm spring contact surface to friction surface should not be less than 52.85 mm (2.081 inches) for 2750 models, 62.85 mm (2.474 inches) for 2755, 2855N and 2955 models.

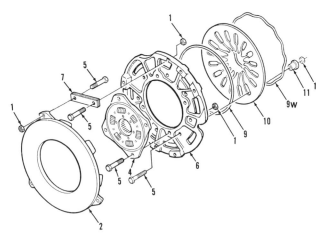

Fig. 100—Exploded view and cross section of LUK 320 mm diameter, single stage, diaphragm-type engine clutch assembly. Some differences may be noted between early style and later style.

1.	Nuts	7.	Plate springs
2.	Pressure plate	9.	Wire ring
4.	Torsion damper	9W.	Waved ring
5.	Screws	10.	Diaphragm spring
6.	Clutch cover	11.	Spacer sleeves

New engine clutch friction disc is 10 mm (0.390 inch) thick. Install new disc if total thickness at facing is less than 7 mm (0.260 inch). When installing, make sure side of disc marked FLYWHEEL SIDE (side with star shaped plate around hub) is toward flywheel. Use a centering tool when installing. Remove centering tool after clutch retaining hex nuts and cap screws are tightened to a torque of 50 N·m (35 ft.-lbs.).

CLUTCH RELEASE BEARING AND SHAFT

All Models Except 2755, 2855N and 2955 with Sound Gard Body

101. REMOVE AND REINSTALL. Separate engine from clutch housing as outlined in paragraph 99. On models so equipped, remove lube tube and fittings. Slide release bearing from carrier. Unbolt fork from release shaft. On models so equipped, disconnect mechanical linkage rod from shaft arm. On models with hydraulic clutch linkage, remove rubber boot from clutch shaft arm. Slide release shaft out left side of clutch housing on all models.

Inspect and renew release shaft bushings as necessary. Drive plastic coated bushings in until flush with clutch housing. **Do not lubricate plastic coated bushings with graphite grease.**

Reinstall components by reversing removal procedure. On models with mechanical clutch linkage, hook short leg of return spring against clutch fork and long leg against clutch housing. Apply Loctite 271 to fork retaining cap screws and tighten them to a torque of 50 N·m (35 ft.-lbs.).

2755, 2855N and 2955 Models with Sound Gard Body

101A. REMOVE AND REINSTALL. Separate engine from clutch housing as outlined in paragraph 99. Refer to paragraph 98 for service to the clutch throwout bearing and hydraulic operating system used on these models.

CLUTCH HOUSING

All Models

102. Clutch housing normally will not need complete removal for servicing. Clutch control linkage can be serviced when clutch housing is separated from engine. Clutch shaft and pto shaft along with their bearings and oil seals and the transmission oil pump can be serviced after clutch housing is separated from transmission case.

HI-LO SHIFT UNIT

OPERATION

All Models with Hi-Lo Shift

103. The Hi-Lo shift unit is a hydraulically shifted gear reduction unit that can be shifted under load in any transmission gear, while on the go, without declutching. Shifting the Hi-Lo unit from Hi to Lo slows tractor ground speed approximately 20 percent and at the same time increases torque to tractor rear axle. Shifting back to Hi, resumes normal ground travel speed.

The Hi-Lo unit is located directly in front of the transmission in the rear of clutch housing.

The transmission oil pump supplies lubricating and pressurized oil to the Hi-Lo unit as well as the front-wheel drive clutch, the pto clutch and is also used as a charge pump for the main hydraulic system pump. Oil from the transmission pump passes through the transmission oil filter, then to the pressure regulating valve, automatic shift valve and Hi-Lo control valve in the transmission shift cover. The transmission oil pressure indicator light should come on when pressure drops to less than about 930 kPa (135 psi) and should go off when pressure increases to about 965 kPa (140 psi). The Hi-Lo shift valve is designed so that, if there is a drop in system oil pressure below 750-850 kPa (109-116 psi), there will be an automatic shift to Lo to prevent slippage and possible damage to the Hi Clutch discs (1 and 2—Fig. 103).

When Hi-Lo control lever is shifted to Hi, pressure oil flows behind Hi Clutch piston (24) and Lo Brake piston (35). Piston (24) overcomes the pressure of Belleville springs (25) and applies Hi Clutch (1 and 2). Brake piston (35) overcomes the pressure of Belleville springs (16) freeing up Lo Brake discs (37 and 39). This allows the drive shaft, sun gear (5), planet carrier, output sun gear (4) and transmission input shaft (42) to rotate as a single unit at engine speed for direct drive.

When control lever is shifted to Lo, pressure oil is removed from behind pistons (24 and 35). Belleville springs (25) force piston (24) back to release Hi Clutch discs (1 and 2). Belleville springs (16) force piston (35) back and applies pressure to Lo Brake discs (37 and 39). Power now flows through unit via drive shaft, input sun gear (5), planet gears (13), output sun gear (4) and transmission input shaft (42).

NOTE: When operating in Hi, if operating pressure drops below 750-850 kPa (109-116 psi) or engine is shut off, Hi-Lo unit will automatically shift to Lo speed position.

TROUBLE-SHOOTING

All Models with Hi-Lo Shift

104. Some problems that may occur during operation of the Hi-Lo unit and their possible causes are as follows:

1. Hi-Lo control lever jumping out of engagement. Could be caused by:
   a. Clogged transmission oil filter.
   b. Low transmission oil pressure.
   c. Broken pressure regulating spring.

2. Transmission oil pressure indicator light glows. Could be caused by:
   a. Faulty transmission oil pump.
   b. Low transmission oil level.
   c. Low setting on pressure regulating valve.
   d. Clogged transmission oil filter.

3. Hi-Lo shifting speed too slow. Could be caused by:
   a. Low transmission oil pressure.
   b. Worn disc pack.
   c. Mechanical failure in Hi-Lo unit.

4. Control lever shifts below specified pressure or does not shift to Hi. Could be caused by:
   a. Faulty gage used for checking pressure.
   b. Stuck automatic shift valve or piston.
   c. Improperly installed check valve at front of down shift valve .

5. Control lever shifts above specified pressure. Could be caused by:
   a. Faulty gage used for checking pressure.
   b. Incorrect assembly of automatic shift valve.
   c. Broken or weak spring at automatic shift valve or piston.
   d. Stuck automatic shift valve or piston .
   e. Missing pin that connects Hi-Lo shift valve and automatic shift piston.

PRESSURE TEST

All Models with Hi-Lo Shift

105. SYSTEM PRESSURE. To check system operating pressure, remove plug (42—Fig. 101) and install a 2070 kPa (300 psi) test gage. Place range and gear shift levers in neutral and apply hand brake. Operate engine at 1500 rpm and warm transmission oil to 65° C (150° F). Operating pressure should be

1050 kPa (150 psi). Adjust pressure, if necessary, by adding or removing shims (64—Fig. 104).

Fig. 101—Hi-Lo system pressure is checked at port where plug (42) is installed. Pressure regulating valve plug is shown at (1). Refer to text.

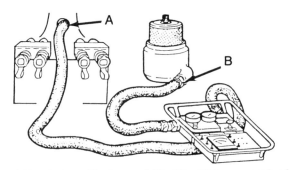

Fig. 102—Automatic down shift of the Hi-Lo control valve can be checked with flow-pressure meter as described in text.

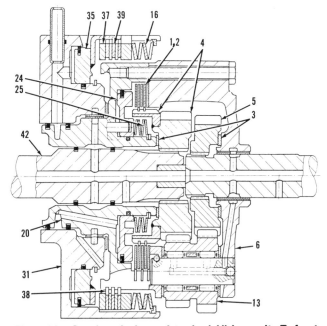

Fig. 103—Sectional view of typical Hi-Lo unit. Refer to Fig. 106 for legend.

106. AUTOMATIC DOWN SHIFT PRESSURE. To check operation of the Hi-Lo automatic down shift and the transmission oil pressure indicator light, attach the hose to a flow and pressure meter to the ported filter cover port (B—Fig. 102), then direct the outlet hose from flow and pressure meter into the transmission filler opening (A). Close the meter control valve, operate engine at 1500 rpm and move Hi-Lo shift handle to "Hi" position. Slowly open the

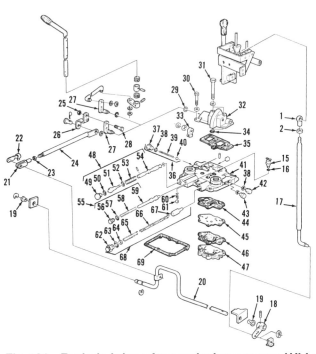

Fig. 104—Exploded view of transmission cover and Hi-Lo control valve typical of that used. Some items are used only on remote shifting for models equipped with Sound Gard Body.

1. Ball socket		
2. Lock nut	42. Plug	
15. Oil pressure switch	43. Plug	
16. "O" ring	44. Gasket	
17. Connecting rod	45. Plate	
18. Lever	46. Gasket	
19. Bushings	47. Manifold cover	
20. Shaft	48. Hi-Lo control valve	
21. Yoke	49. Plug	
22. Retainer	50. "O" ring	
23. Lock nut	51. Piston	
24. Rod	52. "O" ring	
25. Nut	53. Spring	
26. Lever	54. Spool	
27. Shaft	55. Automatic shift valve	
28. Cap screw	56. Plug	
29. "O" ring	57. "O" ring	
30. Cap screw	58. Spring	
31. Cap screw	59. Valve spool	
32. Housing	60. Spring	
33. Arm	61. Ball	
34. "O" ring	62. Plug	
35. Gasket	63. "O" ring	
36. Detent assy.	64. Shim	
37. Plug	65. Spring (outer)	
38. "O" rings	66. Spring (inner)	
39. Spring	67. Valve plunger	
40. Ball	68. Regulating valve	
41. Cover	69. Gasket	

meter control valve while observing the low pressure indicator light and pressure at which the automatic shift to "Lo" occurs. The transmission oil pressure indicator light should come on when pressure drops to less than about 930 kPa (135 psi). The Hi-Lo shift valve should automatically shift to "Lo" when system oil pressure reaches 750-850 kPa (109-116 psi). If pressures are incorrect, refer to TROUBLE-SHOOT-ING paragraph 104. System pressure will increase after lever moves to "Lo".

The low-pressure indicator light should go off when pressure increases to about 965 kPa (140 psi).

OVERHAUL

All Models with Hi-Lo Shift

107. CONTROL AND PRESSURE REGULAT-ING VALVES. Control valve, automatic down shift valve and pressure regulating valve are located in transmission shifter cover as shown in Fig. 104. Regulating valve can be adjusted without removing shifter cover, but valve unit overhaul can be accomplished only after cover is removed.

On models without Sound Gard Body, unbolt and remove transmission shield by working it up over shift lever boots. Disconnect Hi-Lo shift lever linkage and rear wiring harness. Unbolt and remove shift cover, valves and shift levers.

On models with Sound Gard Body, remove floor mat and platform center section. Disconnect Hi-Lo shift linkage and rear wiring harness.

On all models, unbolt and remove housing (32). Remove detent assembly (36), then remove control valve (48).

> **CAUTION: Plugs (56 and 62) are under spring force, use caution when removing them.**

Remove plug (56) and automatic down shift valve (55). Remove plug (62) and regulating valve (68).

Check springs and valves against the following specifications:

Regulating valve outer spring (65)—
  Free length . . . . 129 mm
    (5.080 in.)
  Test length . . . . 64 mm
    (2.52 in.)
  Test load . . . . 95-120 N
    (22-26 lbs.)
Regulating valve inner spring (66)—
  Free length . . . . 182 mm
    (7.170 in.)
  Test length . . . . 109 mm
    (4.290 in.)
  Test load . . . . 85-105 N
    (19-23 lbs.)

Control valve spring (53)—
  Free length . . . . 71 mm
    (2.800 in.)
  Test length . . . . 43 mm
    (1.700 in.)
  Test load . . . . 220-270 N
    (49-60 lbs.)
Control valve detent spring (39)—
  Free length . . . . 49.5 mm
    (1.950 in.)
  Test length . . . . 36.5 mm
    (1.430 in.)
  Test load . . . . 50-60 N
    (11-13 lbs.)
Automatic down shift spring (58)—
  Free length . . . . 71 mm
    (2.800 in.)
  Test length . . . . 43.5 mm
    (1.710 in.)
  Test load . . . . 55-65 N
    (12-15 lbs.)
Control valve (54)—
  Spool OD . . . . 12.65-12.67 mm
    (0.498-0.499 in.)
  Bore ID . . . . 12.73-12.74 mm
    (0.5013-0.5014 in.)
Control valve piston (51)—
  Piston OD . . . . 20.44-20.46 mm
    (0.804-0.805 in.)
  Bore ID . . . . 20.50-20.53 mm
    (0.807-0.808 in.)
Regulating valve (67)—
  Plunger OD . . . . 15.95-15.97 mm
    (0.628-0.629 in.)
  Bore ID . . . . 16.00-16.01 mm
    (0.629-0.630 in.)
Automatic down shift valve (59)—
  Spool OD . . . . 8.96-8.97 mm
    (0.352-0.353 in.)
  Bore ID . . . . 9.00-9.02 mm
    (0.354-0.355 in.)

All springs are available separately, however if valves or bores are excessively worn or damaged, shift cover (41) with valves must be renewed.

Renew "O" rings and gaskets and reassemble and reinstall by reversing disassembly and removal procedures.

108. HI-LO DRIVE UNIT. To remove the Hi-Lo unit, first remove the Sound Gard Body as outlined in paragraph 174, if so equipped; then, on all models, split tractor between clutch housing and transmission as outlined in paragraph 123.

Pull transmission input shaft (1—Fig. 105) out of Hi-Lo unit. Remove oil lines (3, 4 and 7), then remove cap screws (2) and lift out Hi-Lo unit.

Set unit on a bench so that brake housing (31—Fig. 106) is facing downward. Lift planetary with Hi Clutch assembly out of brake housing and lay assembly aside. Use spring compressor JDT-24A and adapter JDT-24-2 or equivalent to compress Belleville springs (16) and remove snap ring (17). Release pressure from Belleville springs, remove springs, release plate (40), brake discs (38) and plates (37 and 39). Remove snap ring (36), piston (35) with piston ring (34) and seal ring (33). Bushing (32) can now be removed if excessively worn or damaged.

Remove seal rings (18) and thrust washer (19) from clutch drum (20). Remove three cap screws (9) and place assembly on a bench with hub of clutch drum (20) facing upward. Remove clutch drum from planet carrier, then remove Hi Clutch plates and discs (1 and 2), thrust washer (3), sun gear and clutch hub (4), input sun gear (5) and thrust washer (3). Push pins (15) upward, remove lock balls (10) and remove pins. Remove planet gears (13) with needle bearings (12), spacers (14) and thrust washers (11). Use spring compressor JDT-24A or equivalent to compress Belleville springs (25), then remove snap ring (26) and springs. Use compressed air, if necessary, to remove piston (24) and piston ring (23). Remove seal ring (22) from clutch drum. Remove bushing (7) from planet carrier if excessively worn or damaged.

Clean and inspect all parts and renew any showing excessive wear or other damage. Check height of Belleville springs against the following specifications:

Brake Belleville spring (16) . . . . 5.61 mm
(0.221 in.)
Clutch Belleville spring (25) . . . . 3.23 mm
(0.127 in.)

Check clutch discs and plates and brake discs and plates against the following new thickness specifications:

Brake back plate (37)—
New . . . . 4.37-4.63 mm
(0.172-0.182 in.)
Brake plate, external tangs (39)—
New . . . . 2.30 mm
(0.091 in.)
Brake disc, internal splines (38)—
New . . . . 2.21-2.37 mm
(0.087-0.093 in.)
Clutch plate, external tangs (1)—
New . . . . 1.50 mm
(0.059 in.)
Clutch disc, internal splines (2)—
New . . . . 2.44-2.54 mm
(0.096-0.100 in.)

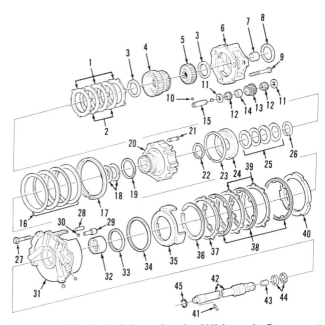

Fig. 106—Exploded view of typical Hi-Lo unit. Some parts may be slightly different than shown.

1. Hi Clutch plates (external tangs)
2. Hi Clutch discs (internal splines)
3. Thrust washers
4. Sun gear & clutch hub
5. Input sun gear
6. Planet carrier
7. Bushing
8. Thrust washer
9. Cap screw (3)
10. Lock ball (3)
11. Thrust washer (6)
12. Needle bearing (6)
13. Planet gear (3)
14. Spacer (3)
15. Pin (3)
16. Lo Brake Belleville springs
17. Snap ring
18. Seal rings
19. Thrust washer
20. Clutch drum
21. Dowel pin (3)
22. Seal ring
23. Piston ring
24. Clutch piston
25. Hi Clutch Belleville springs
26. Snap ring
27. Cap screw (4)
28. Dowel pin (2)
29. Priority valve
30. Valve ball
31. Brake housing
32. Bushing
33. Seal ring
34. Piston ring
35. Brake piston
36. Snap ring
37. Back plate
38. Brake discs (internal tangs)
39. Brake plates (external tangs)
40. Brake release plate
41. Spring pin
42. Transmission input shaft
43. Bushing
44. Seal rings
45. "O" ring

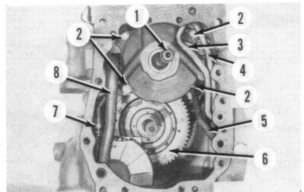

Fig. 105—View into rear of clutch housing, showing Hi-Lo unit, pto clutch and oil lines.

1. Transmission input shaft
2. Cap screws (4)
3. Lubricating line to oil cooler
4. Inlet line from filter
5. Oil pump pressure line
6. Pto clutch
7. Pressure line to front-wheel drive clutch
8. Oil pump suction line

Inside diameter of new bushing (7) is 32.11-32.16 mm (1.264-1.266 inches). Inside diameter of new bushing (32) is 85.27-85.30 mm (3.357-3.358 inches).

Use Fig. 103, Fig. 106 and Fig. 107 as a guide and reassemble by reversing disassembly procedure, keeping the following points in mind: When installing Belleville springs (16), install with concave side toward release plate (40), then alternate concave side up, concave down and concave up toward snap ring (17). When installing clutch piston (24), make certain dowel pin in clutch drum (20) enters hole in piston. When installing Belleville springs (25), start with convex side toward piston (24), then alternate convex side up, convex down and convex up toward snap ring (26). Install clutch plates and discs, alternating clutch plates (1) and internally splined discs (2). To time planet gears, align timing marks on input sun gear (5) with timing marks on planet gears (13). Tighten planet carrier to clutch drum cap screws (9) to a torque of 30 N·m (23 ft.-lbs.).

Reinstall Hi-Lo unit in clutch housing making certain thrust washer (8) is on clutch (drive) shaft.

NOTE: The pto oil inlet line must be guided into the port in lower right front side of brake housing (31) as Hi-Lo unit is being installed.

Use Loctite 271 on cap screws (2—Fig. 105) and tighten to a torque of 50 N·m (35 ft.-lbs.). Install transmission input shaft and connect oil lines. Reconnect tractor and tighten clutch housing to transmission cap screws to 160 N·m (120 ft.-lbs.) torque. Complete installation by reversing removal procedure. Reinstall Sound Gard Body, if so equipped.

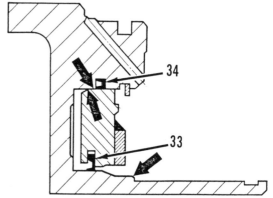

Fig. 107—Cross section showing correct installation of seals (33 and 34). Check for burrs and sharp edges at points indicated by arrows and lubricate with multipurpose grease before assembling.

CREEPER TRANSMISSION

OPERATION

Models with Creeper Transmission

109. The Creeper transmission is a mechanically operated, synchronized reduction drive unit. The 2-speed unit operates as direct drive in Hi position in all transmission speeds and at a 79 percent reduction in Creeper position in Lo range I and reverse. A locking pin (15—Fig. 110) prevents engagement of Creeper drive in Hi range II.

The engine clutch must be disengaged before shifting Creeper transmission. Creeper transmission is located directly in front of the transmission in the clutch housing.

TROUBLE-SHOOTING

Models with Creeper Transmission

110. Some problems that may occur during operation of the Creeper transmission and their possible causes are as follows:

1. Noise during shifts. Could be caused by:
   a. Worn synchronizer discs.
   b. Not fully disengaging engine clutch.

   c. Worn shifter collar.

2. Noise during operation. Could be caused by:
   a. Worn taper roller bearings.
   b. Worn needle bearings.
   c. Incorrect preload of taper roller bearings.

3. Creeper gear jumps out of engagement. Could be caused by:

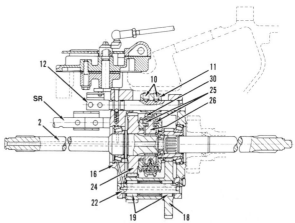

Fig. 108—Cross section of Creeper transmission. Refer to Fig. 110 for legend. Range II shifter rod (SR) is also shown.

a. Worn shifter fork.
b. Incorrectly adjusted shift linkage.
c. Worn or damaged shift linkage.

R&R AND OVERHAUL

Models with Creeper Transmission

111. To remove the Creeper transmission, first remove the Sound Gard Body as outlined in paragraph

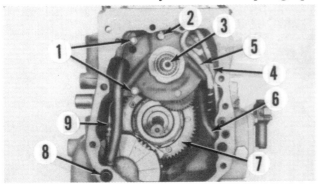

Fig. 109—Rear view of clutch housing showing Creeper transmission and component parts.

1. Creeper retaining cap screws
2. Shifter rod
3. Transmission input shaft
4. Lubricating line to oil cooler
5. Inlet line from filter
6. Oil pump pressure line
7. Pto clutch
8. Sleeve
9. Oil pump suction line

174, if so equipped, then split tractor between clutch housing and transmission as outlined in paragraph 123. Disconnect shift linkage, then unbolt and remove transmission shift cover. Refer to Fig. 109 and remove oil lines (4, 5 and 9) and sleeve (8). Remove cap screws (1) and lift out Creeper transmission.

NOTE: When removing Creeper transmission, do not lose locking pin (15—Fig. 110), which may fall out.

To disassemble the unit, clamp transmission input shaft (2—Fig. 108 or Fig. 110) in a vise with shaft pointing downward. Loosen set screws (10) and remove shifter rod (12) and fork (11). Remove detent ball (14—Fig. 110) and spring (13). Remove Allen screws (39) and lift off end plate (38). Remove bearing cup (36) and shims (37) from end plate. Remove shaft (22) taking care not to lose lock ball (23). Lift out cluster gear (18) with thrust washers (17), needle bearings (19) and spacer (20). Remove input gear (34), pull off bearing cone (35) and press out bearing cup (33). Pull bearing cone (32) from end of transmission shaft (2). Remove snap ring (31) and slide synchronizer assembly (25 through 30) and gear (24) from shaft (2). Remove unit from vise and pull shaft assembly from carrier (16). Discard "O" ring (4) and remove bearing cone (5) from shaft. Remove snap ring (7) and

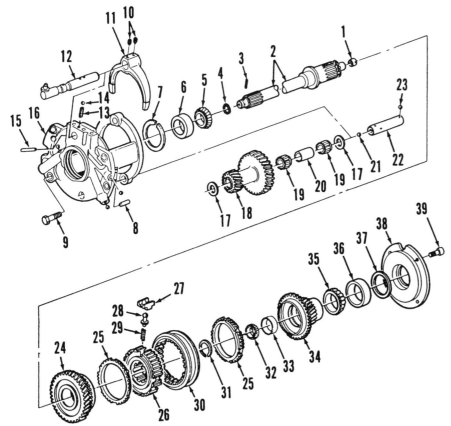

1. Bushing (clutch shaft pilot)
2. Transmission input shaft
3. Spring pin
4. "O" ring
5. Bearing cone
6. Bearing cup
7. Snap ring
8. Dowel pin
9. Cap screw (4)
10. Set screws
11. Shift fork
12. Shifter rod
13. Spring
14. Detent ball
15. Locking pin
16. Carrier
17. Thrust washers
18. Cluster gear
19. Needle bearings
20. Spacer
21. Ball plug
22. Shaft
23. Lock ball
24. Creeper gear
25. Synchronizer discs
26. Synchronizer hub
27. Thrust block (3)
28. Ball pin (3)
29. Spring (3)
30. Shift collar
31. Snap ring
32. Bearing cone
33. Bearing cup
34. Input gear
35. Bearing cone
36. Bearing cup
37. Shim
38. End plate
39. Allen screw (4)

Fig. 110—Exploded view of Creeper transmission assembly.

bearing cup (6) from carrier (16). Disassemble synchronizer unit.

Clean and inspect all parts for excessive wear or other damage and renew as necessary.

Reassemble by reversing disassembly procedure keeping the following points in mind: Refer to Fig. 111 when reassembling the synchronizer. To adjust taper bearing preload, install bearing cup (36—Fig. 110) without shims (37) in end plate (38). Install end plate and secure with Allen screws. Using a dial indicator, check end play of input gear (34). This measurement plus 0.05 mm (0.002 inch) is the correct shim pack (37) to be installed. Remove end plate, press out bearing cup, install shim pack and reinstall bearing cup. Install end plate, apply Loctite 271 to threads of Allen screws and tighten screws to a torque of 25 N·m (20 ft.-lbs.). Tighten set screws (10) to 40 N·m (30 ft.-lbs.) torque.

> NOTE: Before installing Creeper transmission, coat locking pin (15) with grease and install in carrier (16). When joining tractor, make certain that locking pin does not fall out.

Reinstall Creeper transmission assembly, tightening retaining screws to 50 N·m (35 ft.-lbs.) torque. Install oil lines (4, 5 and 9—Fig. 109) and sleeve (8). Install new "O" ring (4—Fig. 110) on transmission input shaft (2). Rejoin tractor and tighten cap screws to 160 N·m (120 ft.-lbs.) torque. Install transmission shift cover and connect shift linkage (Fig. 112). Install Sound Gard Body, if so equipped.

Fig. 111—Install synchronizer hub (26) so that thrust blocks (27) engage the third tooth (arrow) of shift collar as shown.

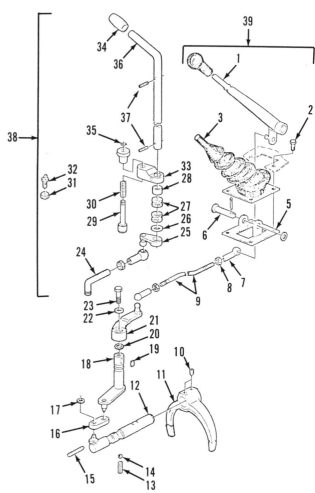

Fig. 112—Exploded view of Creeper transmission shift linkage. Some parts are also shown in Fig. 110.

1.	Control lever	22.	Washer
2.	Cap screw (4)	23.	Cap screw
3.	Cover	24.	Rod
5.	Plate	25.	Arm
6.	Pin	26.	Washer
7.	Ball joint	27.	Bushings
8.	Jam nut	28.	Sleeve
9.	Rod	29.	Pin
10.	Set screw (2)	30.	Spring
11.	Shift fork	31.	Nut
12.	Shifter rod	32.	Screw
13.	Spring	33.	Link
14.	Detent ball	34.	Knob
15.	Locking pin	35.	Snap ring
16.	Link	36.	Control lever
17.	Lock ring	37.	Spring pins
18.	Shifter shaft	38.	Tractors without
19.	Key		Sound Gard Body
20.	"O" ring	39.	Tractors with
21.	Arm		Sound Gard Body

TRANSMISSION
(COLLAR SHIFT)

OPERATION

2750 and 2755 Models with Collar Shift Transmission

112. A collar shift transmission unit with helical cut gears is available on some 2750 and 2755 models. The two control levers are located on top of the clutch housing in shift cover. The left lever selects high or low range and reverse. The right lever selects one of four stepped gears. These shift levers combine to make possible 8 forward and 4 reverse gears.

Disengage engine clutch when shifting transmission gears.

TOP (SHIFT) COVER

2750 and 2755 Models with Collar Shift Transmission

113. REMOVE AND REINSTALL. To remove the shifter cover (1—Fig. 113), first remove transmission shield. Disconnect the hydraulic oil reservoir leak-off and breather line from cover. Detach wire from transmission oil pressure warning switch, then lay wire away from cover. Disconnect Hi-Lo clutch linkage, then unbolt and remove control cover from shift cover. Remove attaching screws and lift shift cover with shift levers from housing. Any further disassembly will be obvious after an examination of the unit and reference to Fig. 113.

Reinstall by reversing the removal procedure.

NOTE: Shifter shafts and forks are an integral part of the transmission and can be serviced after transmission is split from clutch housing as outlined in paragraph 114.

TRACTOR SPLIT

2750 and 2755 Models with Collar Shift Transmission

114. To split tractor between clutch housing and transmission case, first drain transmission case and remove the top shift cover as outlined in paragraph 113. Remove hydraulic oil filter cover and element, then disconnect all wires and tubes that would interfere when separating. Support rear of tractor under transmission case and front of tractor under clutch housing.

CAUTION: Be sure that tractor is supported adequately to permit separation, but not to allow tipping or other unwanted movement.

Remove the two clutch housing to transmission case cap screws at rear of shift cover opening under the mounting flange. Remove the remaining nine screws, then separate clutch housing and transmission case.

NOTE: Do not lose the pto shaft coupling. If the pto coupling falls off shaft and into transmission

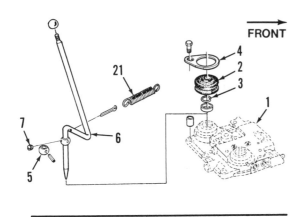

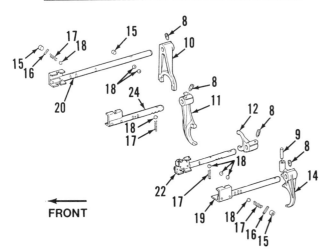

Fig. 113—Exploded view of shift cover, shifter shafts and shift forks used.

1. Shift cover
2. Boot
3. Snap ring
4. Retainer
5. Bushing
6. Shift lever
7. Set screw
8. Set screws
9. Start safety switch pin
10. Fork (1-5 & 2-6)
11. Fork (3-7 & 4-8)
12. Fork (Range II)
14. Fork (Range I-Reverse)
15. Plug
16. Spring pins
17. Detent springs
18. Detent balls
19. Shifter shaft (Range I-Reverse)
20. Shift shaft (1-5 & 2-6)
21. Spring
22. Shifter shaft (Range II)
24. Shift shaft (3-7 & 4-8)

case, use screwdriver to reposition coupling on rear shaft before attempting to reattach clutch housing and transmission case.

When reassembling, move clutch housing and transmission case together completely before installing attaching cap screws. Coat threads of screw with sealer before installing in hole third from top on right. Remainder of assembly is reverse of disassembly procedure. Tighten clutch housing to transmission case cap screws to a torque of 160 N·m (120 ft.-lbs.).

SHIFTER SHAFTS AND FORKS

2750 and 2755 Models with Collar Shift Transmission

115. REMOVE AND REINSTALL. To remove shifter shafts and forks, first split tractor as outlined in paragraph 114. Remove seat and disconnect lift links from rockshaft arms. Detach rear wiring harness and disconnect lines from selective control valves to rockshaft housing. Attach a hoist to rockshaft housing, place load selector lever in "L" position, then unbolt and remove rockshaft housing from transmission case.

Remove starter safety switch pin (9—Fig. 113) and set screws (8) from shift forks (12 and 14).

Move both range shifter shafts (19 and 22—Fig. 114) to neutral position, then pull shaft (19) out of case and shift fork (14). **Do not turn shifter shafts while withdrawing from housing. Detent ball could drop into set screw hole, making shaft impossible to remove. Also, do not lose interlock balls, detent balls or springs when shifter shafts are removed from case bores.** Pull shaft (22) from case and shift fork (12). Remove interlock balls, detent balls and springs.

Move shifter shafts (20 and 24) to neutral position, remove set screws from shift forks (10 and 11), then withdraw shifter shaft (24) from case bore. Carefully remove interlock balls, detent balls and springs. Remove shifter shaft (20). Remove forks (10 and 11) as shafts are withdrawn.

Inspect all parts for evidence of wear, rust or other damage and renew as necessary. Assemble shifter shafts and forks in reverse of order in which they were removed.

COUNTERSHAFT

2750 and 2755 Models with Collar Shift Transmission

116. R&R AND OVERHAUL. To remove the countershaft, split tractor as outlined in paragraph 114 and remove shift forks and shaft as outlined in

paragraph 115. Remove the pto gear, or gears, from front of transmission. Remove cap screws from countershaft bearing support (13—Fig. 116). Remove snap ring (33) from its groove at rear end of countershaft, then use a screwdriver to turn lock washer (29) until splines of washer index with splines of countershaft. Pry bearing support off dowels, pull assembly forward, notice that shoulder on shift collar (28—Fig. 115 or Fig. 116) is toward front and lift gears from transmission case as they come off shaft.

Inspect all gears, thrust washers and shift collar for broken teeth, excessive wear or other damage and renew as necessary. If support assembly bearings or shafts require service, the shafts and bearings can be pressed out after removing snap rings; however, the transmission drive gear must be removed before countershaft can be removed. Snubber brake springs (22—Fig. 116) should test 280-350 N (63-79 lbs.) when compressed to a height of 38.5 mm (1.51 inches). Needle bearing (35) can be removed from its bore after removing snap ring (34).

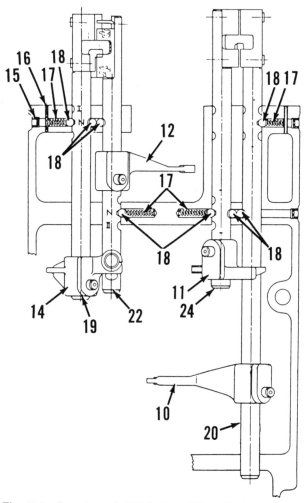

Fig. 114—Drawing of shift forks, shifter shafts and associated parts. Refer to Fig. 113 for legend.

INPUT SHAFT

2750 and 2755 Models with Collar Shift Transmission

117. R&R AND OVERHAUL. To remove the input shaft, it is first necessary to remove the countershaft as outlined in paragraph 116.

With countershaft removed, the transmission input shaft is removed as follows: Remove transmission oil

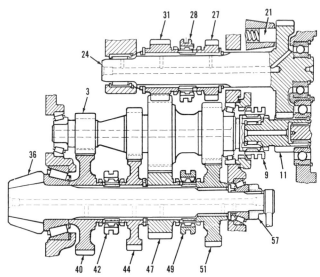

Fig. 115—Cross section of transmission shafts, gears and collars. Refer to Fig. 116 for legend.

cup and lines. Straighten lock plates and remove input shaft bearing quill (8—Fig. 116) and shims (7) from front of input shaft. Bump input shaft forward, lift rear end of shaft and remove input shaft from transmission case.

With input shaft removed, inspect all gears for chipped teeth or excessive wear. Inspect bearings and renew as necessary. Bump bearing cup (1) forward if removal is required. Inspect needle bearing (4) and renew if necessary.

Install and adjust end play of input shaft as follows: Be sure bearing cup (1) is bottomed in bore and place input shaft in position. Use original shim pack (7), or use a shim pack approximately 0.8 mm (0.030 inch) thick, install front bearing quill (8) and tighten retaining cap screws to 50 N•m (35 ft.-lbs.) torque. Use a dial indicator (D—Fig. 117) to check input shaft end play, which should be 0.10-0.15 mm (0.004-0.006 inch). Vary thickness of shims (7—Fig. 116) as required. Do not forget front oil line clamp when making final assembly.

PINION SHAFT

2750 and 2755 Models with Collar Shift Transmission

118. R&R AND OVERHAUL. To remove the transmission pinion shaft, first remove the shift forks

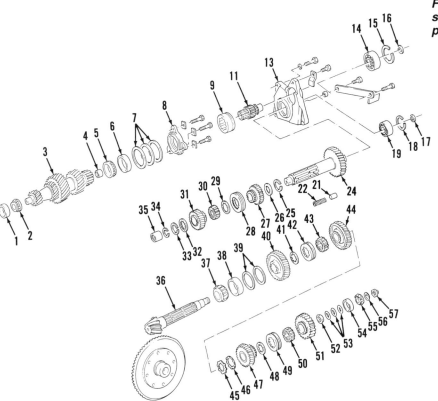

Fig. 116—Exploded view of transmission shafts, gears, shift collars and associated parts used on collar shift units.

1. Bearing cup	32. Thrust washer
2. Bearing cone	33. Snap ring
3. Input shaft	34. Snap ring
4. Needle bearing	35. Needle bearing
5. Bearing cone	36. Pinion shaft
6. Bearing cup	37. Bearing cone
7. Shims	38. Bearing cup
8. Bearing quill	39. Shims
9. Shifter collar	40. 1st & 5th gear
11. Drive gear	41. Thrust washer
13. Support	42. Shift collar
14. Ball bearing	43. Shift collar sleeve
15. Snap ring	44. 2nd & 6th gear
16. Snap ring	45. Thrust washer
17. Snap ring	(outer tangs)
18. Snap ring	46. Retaining washer
19. Ball bearing	47. 4th & 8th gear
21. Brake plug	48. Thrust washer
22. Spring	49. Shift collar
24. Countershaft	50. Shift collar sleeve
25. Snap ring	51. 3rd & 7th gear
26. Thrust washer	52. Spacer
27. Reverse pinion	53. Shims
28. Shift collar	54. Bearing cup
29. Thrust washer	55. Bearing cone
(locking)	56. Special washer
30. Shift collar sleeve	57. Nut
31. Low range pinion	

and shafts as outlined in paragraph 115, the countershaft as outlined in paragraph 116, the input shaft as outlined in paragraph 117, both final drive assemblies as outlined in paragraph 135 and the differential as in paragraph 129.

Remove oil line, nut (57—Fig. 116), bearing (55), shims (53) and spacer (52). Use screwdriver and turn thrust washers (41, 45 and 48) until splines of thrust washers are indexed with splines of pinion shaft. Pull pinion shaft rearward and remove parts from transmission case as they come off shaft. Bearing cup (38) and shims (39) can be removed from housing by bumping cup rearward. Be sure to keep shims (39) together as they control the bevel gear mesh position. Bearing cup (54) can be removed by bumping cup forward.

Check all gears and shafts for chipped teeth, damaged splines, excessive wear or other damage and renew as necessary. If pinion shaft is renewed, it will also be necessary to renew the differential ring gear and right hand differential housing as these parts are not available separately. Bearing (37) is installed with large diameter toward gear end of shaft.

NOTE: Mesh (cone point) position of the pinion shaft and main drive bevel pinion gear is adjusted with shims (39) located between rear bearing cup (38) and housing. If new drive gears or bearings are installed, the mesh position must first be checked and adjusted as outlined in paragraph 132.

Install pinion shaft and adjust shaft bearing preload as follows: Use Fig. 116 as a guide and with bearing (37) on pinion shaft, start shaft into rear of housing. With shaft about halfway into housing, place 1st and 5th speed gear (40) on shaft with teeth for shift collar (42) toward front. Place thrust washer (41) on shaft, then install coupling sleeve (43) and shift collar (42). Move shaft forward slightly and install 2nd and 6th speed gear (44) with teeth for shift collar toward rear. Place thrust washer with outer tangs (45) over shaft, then slide retaining washer (46) over thrust washer (45). Move shaft slightly forward and install 4th and 8th speed gear (47) on shaft with teeth for shift collar toward front. Place thrust washer (48) on shaft and install shift collar sleeve (50) and shift collar (49). Install 3rd and 7th speed gear (51) on shaft with teeth for shift collar toward rear. Push shaft forward until rear bearing cone (37) seats in bearing cup (38) and use screwdriver to turn thrust washer until splines on thrust washers lock with splines of pinion shaft. Install spacer (52), shims (53), bearing (55), washer (56) with tang facing forward, and nut (57), then adjust pinion shaft bearing preload as outlined in paragraph 119.

119. PINION SHAFT BEARING ADJUSTMENT. The pinion shaft bearings must be adjusted

to provide a bearing preload of 0.15 mm (0.006 inch). Adjustment is made by varying the number of shims (53—Fig. 116).

To adjust the pinion shaft bearing preload, proceed as follows: Mount a dial indicator (D—Fig. 118) with contact button on front end of pinion shaft and check for shaft end play. If shaft has no end play, add shims (53—Fig. 116) to introduce not more than 0.05 mm (0.002 inch) shaft end play.

NOTE: Do not exceed 0.05 mm (0.002 inch) shaft end play when beginning adjustment because more end play increases the probability of inaccurate measurements and resulting improper adjustment of an important bearing preload adjustment.

If original shims (53) are not being used, install preliminary 0.85 mm (0.035 inch) thickness shim pack. Shims are available in 0.05, 0.13, 0.25 and 0.80 mm (0.002, 0.005, 0.010 and 0.032 inch) thicknesses. Tighten nut (57) to 220 N·m (160 ft.-lbs.) torque and measure shaft end play. Remove shims equal to the measured end play PLUS an additional 0.15 mm (0.006 inch). This will give the recommended bearing preload of 0.15 mm (0.006 inch). Retighten nut (57) to 220 N·m (160 ft.-lbs.) torque and stake in position.

Fig. 117—Transmission drive (input) shaft end play can be measured with dial indicator (D) as shown. Adjust by adding or removing shims (7—Fig. 116).

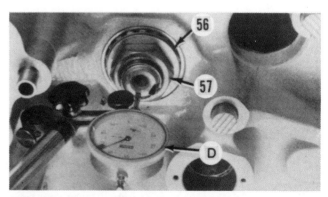

Fig. 118—Pinion shaft end play can be measured with a dial indicator (D). Adjust by adding or removing shims (53—Fig. 116).

TRANSMISSION OIL PUMP

2750 and 2755 Models with Collar Shift Transmission

The transmission oil pump used is an internal gear, crescent-type pump located in the clutch housing and driven by the pto clutch drive shaft. Oil from the transmission oil pump is delivered to the oil cooler, to tractor brakes and to the lubricating system for the collar shift transmission. The transmission oil pump also delivers oil to the main hydraulic pump.

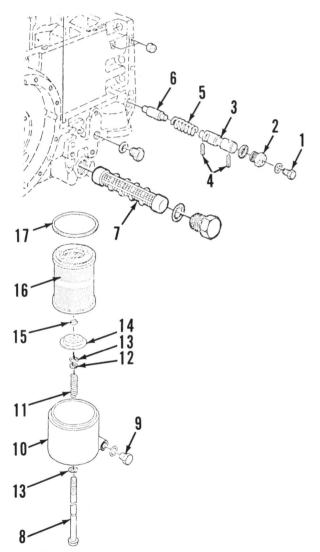

Fig. 119—View showing locations of transmission oil pump screen (7), filter (16) and bypass valve (3).

1. Plug	
2. Threaded nipple	10. Filter cover
3. Bypass valve	11. Compression spring
4. Spring pins	12. Washer
5. Spring	13. "O" ring
6. Pin	14. Retainer
7. Inlet screen	15. Snap ring
8. Bolt	16. Filter
9. Plug	17. Seal washer

120. A bypass valve (1 through 6—Fig. 119) is located above the filter in transmission housing. If the filter (16) becomes clogged or if oil is too cold to circulate through filter easily, the differential pressure (before the filter and after the filter) will exceed the opening pressure of the bypass valve and oil will be directed back to the main oil reservoir (transmission case). Differential pressure should be 200-350 kPa (28.5-50 psi). Specifications for spring (5) should be as follows:

Model 2750
Free length.................... Approx. 68 mm
(2.68 in.)
Pressure when compressed to 49 mm...... 50-60 N
(1.5 in.)   (11-13 lbs.)
Pressure when compressed to 38 mm..... 80-95 N
(1.93 in.)   (18-21 lbs.)

Model 2755
Free length.................... Approx. 65 mm
(2.56 in.)
Pressure when compressed to 33 mm..... 42-52 N
(1.3 in.) (9.5-11.5 lbs.)

If flow volume is suspected to be low, remove, clean and inspect transmission pump inlet screen (7). If screen is clean, separate tractor between clutch housing and transmission case and check the following:
a. Pump inlet and discharge tube "O" rings and packings for deterioration.
b. Pump body screws for proper torque.
c. Pump body and gears for excessive wear or scoring.

121. R&R AND OVERHAUL. To remove the transmission oil pump, first split tractor between clutch housing and transmission as outlined in paragraph 114.

With clutch housing separated from transmission case, pull pto clutch from clutch housing. Remove oil pump suction and pressure lines and retainers. Remove the three oil pump retaining cap screws, then pull oil pump with pto clutch drive shaft out of clutch housing.

Remove cap screws (6—Fig. 120) and separate pto clutch drive shaft (4) from pump assembly. Remove snap ring (9) and press bearing (8) from shaft. Remove snap ring (1), then using a slide hammer puller, remove needle bearing (2), oil seal (3) and cap (3A). Remove cap screws (10) and separate bearing housing (12) from pump housing and carrier (17). Remove spur gear (15) and internal gear (14) from pump housing and carrier. Remove oil seal (16) and "O" ring (18).

Clean and inspect all parts and renew any showing excessive wear or other damage. Bearing housing

(12), internal gear (14), spur gear (15) and pump housing and carrier (17) are available only as an assembly.

Reassemble oil pump by reversing disassembly procedure, keeping the following points in mind: Install gears (14 and 15) in pump housing and carrier (17), chamfered side first. Place a straightedge across sealing surface of housing and carrier and measure clearance between gears and straightedge. Clearance should be 0.03-0.05 mm (0.001-0.002 inch). Clean sealing surfaces of pump housing and carrier (17) and bearing housing (12) with Loctite primer, then coat with Loctite 515. Tighten cap screws (6 and 10) to a torque of 55 N·m (40 ft.-lbs.). Fill pump with transmission oil and check pump for free rotation. Coat sealing surface of clutch housing partition with Loctite 515 and coat carrier sleeve with Moly High Temperature EP grease. Install oil pump assembly and tighten cap screws to a torque of 55 N·m (40 ft.-lbs.).

The balance of reassembly is the reverse of disassembly. Refer to paragraph 114 and rejoin tractor.

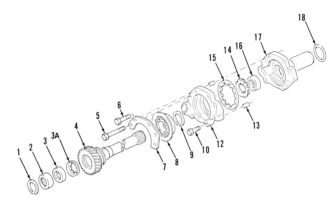

Fig. 120—Exploded view of transmission oil pump used on models with collar shift transmission.

1. Snap ring	9. Snap ring
2. Needle bearing	10. Cap screw (2)
3. Oil seal	12. Bearing housing
3A. Cap (late models)	13. Dowel pins (2)
4. Pto clutch drive shaft	14. Internal gear
5. Cap screw (3)	15. Spur gear
6. Cap screw (2)	16. Oil seal
7. Retainer	17. Pump housing & carrier
8. Ball bearing	18. "O" ring

SYNCHRONIZED TRANSMISSION

OPERATION

All Models with Synchronized Transmission

122. The transmission used on some models is of the cone synchronized type. Tractors without Sound Gard Body are equipped with center shift (two shift levers in shift cover). Tractors with Sound Gard Body are equipped with console shift (two shift levers in console to right of operator's seat). The range shift lever (left lever on center shift or right lever on console shift) is used to select low range I, high range II or reverse range. The gear shift lever (right lever on center shift or left lever on console shift) is used to select the four synchronized gears of each range. The transmission, when combined with the Hi-Lo shift unit, provides a total of 16 forward gears and eight reverse gears.

Tractors are equipped with a neutral start safety switch, which requires that Range shift lever be in neutral position before engine can be started.

TRACTOR SPLIT

All Models with Synchronized Transmission

123. To split tractor between clutch housing and transmission case, drain transmission and disconnect battery cables. On models so equipped, remove Sound Gard Body as outlined in paragraph 174. On models equipped with front-wheel drive, disconnect front drive shaft. If so equipped, disconnect Hi-Lo or Creeper shift linkage. On all models, disconnect hydraulic reservoir leak-off and breather line from shifter cover. Detach wire from transmission oil pressure warning switch. Remove attaching cap screws and remove shifter cover assembly. Disconnect all wires and hydraulic tubes that would interfere when separating tractor. Support rear of tractor under transmission case and front of tractor under clutch housing. Place wood blocks between front axle and front support to prevent tipping. Remove the two clutch housing to transmission case cap screws located at rear of shifter cover opening. Remove the remaining cap screws, then separate clutch housing and transmission case.

NOTE: On models equipped with Creeper transmission, be careful not to lose locking pin (15—Fig. 110) which may fall out when splitting tractor. Coat locking pin with grease and install in carrier (16) when rejoining tractor.

Rejoin tractor and tighten cap screws to a torque of 160 N·m (120 ft.-lbs.). Balance of reassembly is the reverse of disassembly. Fill transmission to top mark on dipstick with John Deere Hy-Gard or Quatrol oil or equivalent. On models so equipped, install Sound Gard Body.

SHIFT LINKAGE, SHIFTER SHAFTS AND FORKS

All Models with Center Shift Synchronized Transmission

124. REMOVE AND REINSTALL. To remove the shift levers (2—Fig. 121) and shift cover (9), first remove transmission shield by working it up over shift levers. Disconnect Hi-Lo or Creeper shift linkage, if so equipped. Disconnect rear wiring harness, then unbolt and remove shift cover assembly. To remove shifter shafts and shift forks, first split tractor between clutch housing and transmission case as outlined in paragraph 123. Remove rockshaft assembly as in paragraph 166. Remove safety start switch pin (12—Fig. 121). Remove set screws (10) from shift forks and guide. Place shifter shafts in neutral position, then pull shifter shafts one at a time forward and remove. Be careful to catch detent and interlock balls as shafts are withdrawn.

Remove shift forks and guide. To remove low range I and reverse shift fork, it is necessary to remove transmission bearing quill (70—Fig. 124).

Clean and inspect all parts and renew any showing excessive wear or other damage.

Reassemble by reversing disassembly procedure. Make certain detent balls (17—Fig. 121), springs (18) and interlock balls (19) are installed correctly. Adjust position of shift forks before set screws (10) are fully tightened. Shift forks must be positioned on shifter shafts so that shift forks do not rub against shifter collar. Set position of shift rails to neutral detent position and tighten set screws lightly. Check position of shift forks in synchronizer collars and make sure that all are centered in neutral. After forks are correctly adjusted, tighten set screws evenly to a torque of 40 N·m (30 ft.-lbs.). Make sure that forks do not rub against any part of collar.

Reinstall rockshaft assembly and rejoin tractor. Install shift cover and transmission shield.

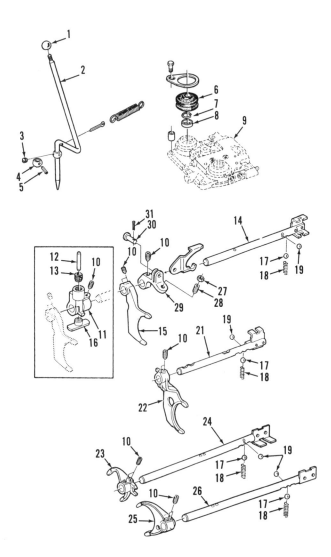

Fig. 121—Exploded view of shift levers, shifter shafts and shift forks used on models with center shift. Guide parts (28, 29, 30 and 31) are used on later models instead of parts (11, 12, 13 and 16).

1. Knob
2. Shift lever (2)
3. Special set screw
4. Ball
5. Spring pin
6. Boot
7. Snap ring
8. Bearing
9. Shift cover
10. Set screws
11. Guide
12. Safety start switch pin
13. Pin bushing
14. Shifter shaft (high range II)
15. Shift fork (high range II)
16. Shoe
17. Detent ball (4)
18. Spring (4)
19. Interlock ball (4)
21. Shifter shaft (low range I & reverse)
22. Shift fork (low range I & reverse)
23. Shift fork (3-7 & 4-8)
24. Shifter shaft (3-7 & 4-8)
25. Shift fork (1-5 & 2-6)
26. Shifter shaft (1-5 & 2-6)
27. Lock nut
28. Adjusting screw
29. Guide
30. Pin
31. Cotter pin

All Models with Console Shift Synchronized Transmission

125. REMOVE AND REINSTALL. To remove transmission shift linkage, refer to Fig. 122 and remove cap screws (4) and rockshaft lever stop nut (5). Remove console side panel (3). Remove knobs from hand throttle, selective control valve, rockshaft control lever, gear shift lever and range shift lever. Disconnect wire at Hi-Lo shift unit indicator light. Unbolt console from fender and lift console (6) upward and remove. Disconnect throttle rod, Hi-Lo connecting rod, if so equipped, and shift rods (14, 15, 16 and 17—Fig. 123). Unbolt and remove bracket (4) with shift lever assemblies. Disconnect and remove shift rods from lever shafts (18, 19, 20 and 21).

To remove shifter shafts and shift forks, first remove Sound Gard Body as outlined in paragraph 174, separate clutch housing from transmission case as in paragraph 123 and remove rockshaft assembly as in paragraph 166. Move all shifter shafts to neutral position. Mark position of shift arms (28, 29, 35 and 36—Fig. 123) in relation to lever shafts (18, 19, 20 and 21). Loosen set screws in shift arms, pull lever shafts from housing (24) and remove shift arms. Loosen set screws in shift forks, safety start switch arm (33) and guide (38). Pull shifter shafts (34 and 37), one at a time, out front of transmission and remove shift fork (32), safety start switch arm (33) and guide (38). Catch detent and interlock balls as shafts are removed.

CAUTION: Do not turn shifter shafts when pulling them forward as detent balls and springs may fall out.

Pull shifter shafts (27 and 31), one at a time, out front of transmission and remove forks (26 and 30). To remove shift fork (39), it is necessary to remove

bearing quill (70—Fig. 124) from front of transmission. Unbolt and remove housing (24—Fig. 123).

Clean and inspect all parts and renew any showing excessive wear or other damage.

Reassemble by reversing disassembly procedure. Make certain detent balls (41), springs (42) and interlock balls are installed correctly. Adjust position of

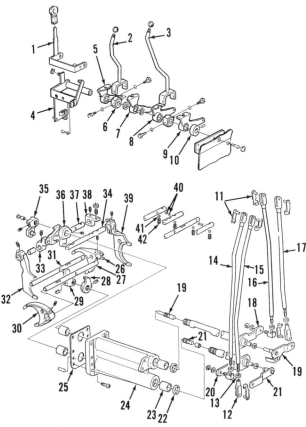

Fig. 123—Exploded view of transmission shift linkage, shifter shafts and shift forks used on models equipped with console shift.

1. Hi-Lo lever	
2. Gear shift lever	24. Housing
3. Range shift lever	25. Gasket
4. Mounting bracket	26. Shift fork (3-7 & 4-8)
5. Shift lever	27. Shifter shaft (1-5 & 2-6)
6. Bushing	28. Shift arm
7. Shift lever	29. Shift arm
8. Shift lever	30. Shift fork (1-5 & 2-6)
9. Shift lever	31. Shifter shaft (3-7 & 4-8)
10. Locating ring	32. Shift fork (high range II)
11. Clip pin	33. Safety start switch arm
12. Yoke	34. Shifter shaft
13. Lock nut	(low range I & reverse)
14. Rod (1-5 & 2-6)	35. Shift arm
15. Rod (3-7 & 4-8)	36. Shift arm
16. Rod (low range I & reverse)	37. Shifter shaft
17. Rod (high range II)	(high range II)
18. Hollow lever shaft (upper)	38. Guide
19. Lever shaft (upper)	39. Shift fork
20. Hollow lever shaft (lower)	(low range I & reverse)
21. Lever shaft (lower)	40. Interlock balls (4)
22. Seal ring (2)	41. Detent ball (4)
23. Bushing (4)	42. Spring (4)

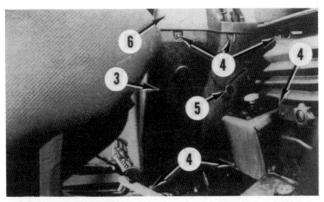

Fig. 122—View of console typical of all models equipped with Sound Gard Body.

3. Side panel	5. Rockshaft lever stop
4. Cap screws	6. Console cover

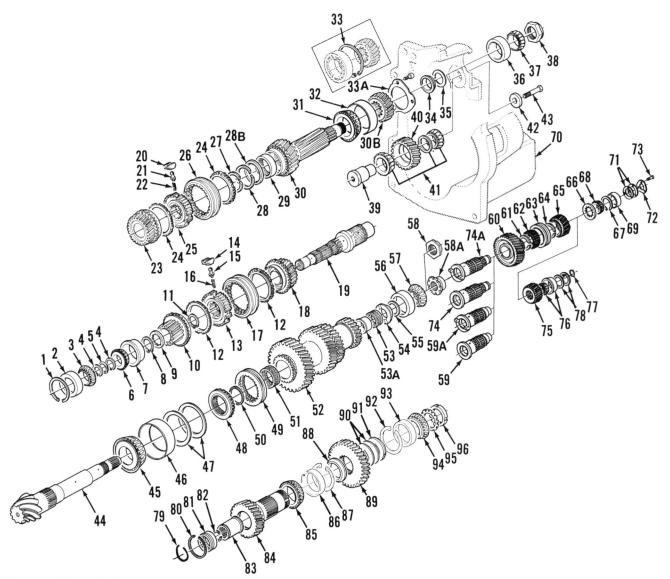

Fig. 124—Exploded view of transmission shafts, gears and associated parts typical of all models with synchronized transmission. Items (74 through 78 and 84 through 96) are used only on models equipped with front-wheel drive and parts with "A" suffix are used on later models. Removable reverse gear (30B) is not used on 2750 models and reverse gear is machined as part of drive shaft (30).

1. Snap ring	24. Synchronizer ring	
2. Bearing cup	25. Synchronizer hub	41. Bearing assy.
3. Bearing cone	26. Shift collar	42. Washer
4. Snap rings	27. Snap ring	43. Cap screw
5. "O" ring	28. Snap ring	44. Bevel pinion shaft
6. Bearing cone	28B. Thrust ring	45. Bearing cone
7. Bearing cup	29. Needle bearing	46. Bearing cup
8. Snap ring	30. Drive shaft	47. Shim
9. Thrust washer	(low range	48. Drive sleeve
10. Gear (1-5)	I-reverse & 4-8)	(range II)
11. Snap ring	30B. Reverse gear	49. Shift collar
12. Synchronizer ring	(some models)	50. Thrust washer
13. Synchronizer hub	31. Bearing cone	51. Needle bearing
14. Thrust block	32. Bearing cup	52. Countershaft
15. Ball pin	33. Snap ring	cluster gear
16. Spring	33A. Bearing retainer	53. Needle bearing
17. Shift collar	34. Spacer	53A. Needle bearing
18. Gear (2-6)	35. Shim	race
19. Drive shaft	36. Bearing cup	54. Spacer
20. Thrust block	37. Bearing cone	55. Shim
21. Ball pin	38. Nut	56. Bearing cup
22. Spring	39. Shaft	57. Bearing cone
23. Gear (3-7)	40. Reverse idler gear	58. Special nut
		58A. Special nut

59. Extension shaft (early models)	75. Gear (front-wheel drive)
59A. Extension shaft (late models)	76. Taper roller bearing
60. Gear (low range I)	77. Snap ring
61. Snap ring	78. Shims
62. Shift sleeve	79. Snap ring
63. Shift collar	80. Snap ring
64. Snap ring	81. Needle bearing
65. Gear (reverse range)	82. Snap ring
66. Thrust washer	83. Pto connecting sleeve
67. Snap ring	84. Intermediate shaft
68. Bearing cone	85. Bearing cone
69. Bearing cup	86. Bearing cup
70. Bearing quill	87. Snap ring
71. Shim	88. Thrust ring
72. Plate	89. Intermediate gear
73. Cap screw (3)	90. Shim
74. Extension shaft (early FWD)	91. Thrust ring
	92. Snap ring
74A. Extension shaft (late FWD)	93. Bearing cup
	94. Bearing cone
	95. Lock ring
	96. Special nut

shift forks before set screws are fully tightened. Shift forks must be positioned on shifter shafts so that shift forks do not rub against shifter collar. Set position of shift rails to neutral detent position and tighten set screws lightly. Check position of shift forks in synchronizer collars and make sure that all are centered in neutral. After forks are correctly adjusted, tighten set screws evenly to a torque of 40 N·m (30 ft.-lbs.). Make sure that forks do not rub against any part of collar.

Reinstall rockshaft assembly and rejoin tractor. Install Sound Gard Body.

SHAFTS AND GEARS

All Models with Synchronized Transmission

126. R&R AND OVERHAUL. To remove the shafts and gears, first separate clutch housing from transmission case as outlined in paragraph 123. Remove rockshaft housing assembly as in paragraph 166 and shifting mechanism as in paragraph 124 or 125. Refer to Fig. 125, straighten locking tab and remove special nut (38). Remove cap screws (A), then remove bearing quill with reverse idler and if equipped with front-wheel drive, intermediate shaft (84). Remove shaft spacer (34—Fig. 124), shim (35) and gear (30B), if so equipped. Remove snap ring (33—Fig. 126) from early models or screws and retainer plate (33A) from later models. Pull extension shaft (74—Fig. 127) with shift fork (A) off bevel pinion shaft. Remove oil cup with lube oil lines. Unbolt and remove bearing cap (B—Fig. 128), then remove snap

ring (1—Fig. 124) from housing groove. Lift top shaft assembly (A—Fig. 128) from top opening of transmission case. Remove differential assembly as outlined in paragraph 129. Remove special nut (58 or 58A—Fig. 124), drive bevel pinion shaft (44) rearward and pull bearing cone (57) from shaft. Remove shim (55) and spacer (54). Pull bevel pinion shaft (44) out rear of transmission case, taking care not to damage needle bearings (51 and 53) in countershaft cluster gear. Remove countershaft cluster gear (52) with needle bearings, thrust washer (50), shift collar (49) and drive sleeve (48). If necessary, remove all bearing cups still located in transmission case. Needle bearing race (53A) can be removed if new race is to be installed. Heat new race before installing.

To disassemble transmission bearing quill, remove cap screw (43), then remove shaft (39), reverse idler gear (40) and bearing assembly (41).

On tractors without front-wheel drive, remove snap ring from groove in pto connector sleeve (83) and pull sleeve out of needle bearing (81). Remove snap rings, then push needle bearing (81) from bearing quill.

On tractors equipped with front-wheel drive, straighten lock ring (95) and remove special nut (96). Drive intermediate shaft (84) inward and remove bearing cone (94). Pull intermediate shaft out of bearing quill. Remove gear (89), thrust rings (88 and 91) and shims (90). Remove snap ring (80) and pull pto connecting sleeve (83) with needle bearing (81) out of intermediate shaft (84). Remove snap ring from con-

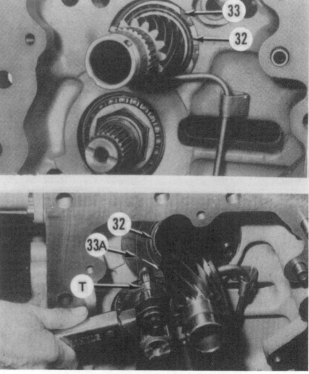

Fig. 126—Early models are equipped with snap ring (33) to retain bearing cup (32) and late models use Torx screws to retain plate (33A).

Fig. 125—Front bearing quill installed on early model equipped with front-wheel drive.

A. Cap screws
30. Low range I-reverse drive shaft
38. Special nut

43. Reverse idler cap screw
74. Extension shaft
84. Intermediate shaft

necting sleeve, then remove needle bearing from sleeve. Remove bearing cone (85) from intermediate shaft. If necessary, remove snap rings (87 and 92) and press bearing cups (86 and 93) from bearing quill.

To disassemble the top shaft assembly, separate low range I-reverse drive shaft (30) from drive shaft (19). Remove snap ring (28), thrust ring (28B) and needle bearing (29) from shaft (30), then pull bearing cone (31) from shaft. Remove snap ring (27) from drive shaft (19) and slide synchronizer (20, 21, 22, 24, 25 and 26) and gear (23) from drive shaft. Pull bearing cone (3) off shaft, then remove snap rings (4) and "O" ring (5). Pull bearing cone (6) from shaft, remove snap

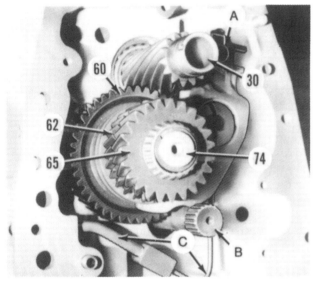

Fig. 127—Front view of transmission with bearing quill removed.

A. Low range I-reverse range
shift fork
B. Pto drive shaft
C. Lube oil line
30. Low range I-reverse
drive shaft

60. Gear (Low range I)
62. Gear (reverse)
65. Drive gear
(front-wheel drive)
74. Extension shaft

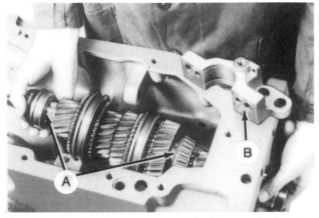

Fig. 128—View showing removal of transmission top shaft assembly.

A. Top shaft assy.
B. Bearing cap

ring (8), thrust ring (9), gear (10) and snap ring (11). Slide synchronizer assembly (12 through 17) and gear (18) from drive shaft (19). Some parts of synchronizers may be alike, but should be identified for installation in the same location.

Disassemble extension shaft as follows: On models without front-wheel drive, remove snap ring (67) and pull bearing cone (68) off extension shaft (59). Remove thrust washer (66) and gear (65) from extension shaft. Remove snap ring (64), shift collar (63), shift sleeve (62), snap ring (61) and gear (60) from shaft (59 or 59A).

On models equipped with front-wheel drive, remove snap ring (77) and pull bearing cone (76) from extension shaft (74 or 74A). Remove gears (75 and 65), snap ring (64), shift collar (63), shift sleeve (62), snap ring (61) and gear (60) from shaft (74 or 74A).

Clean and inspect all parts and renew any showing excessive wear or other damage. Later extension shafts (59A and 74A) have lugs that lock position of nut (58A) when assembling.

NOTE: If bevel pinion shaft is worn or damaged and must be renewed, also renew ring gear and housing assembly as these parts are a matched set.

Reassemble transmission by reversing disassembly procedure and keeping the following points in mind: When installing reverse idler gear and shaft assembly in bearing quill, tighten cap screw (43—Fig. 124) to a torque of 55 N·m (40 ft.-lbs.).

If equipped with front-wheel drive, install intermediate shaft (84) and component parts (86 through 96) in bearing quill, using a shim pack (90) 1.15 mm (0.045 inch) thick. Tighten special nut to a torque of 140 N·m (100 ft.-lbs.). Using a dial indicator, measure intermediate shaft end play. This measurement plus 0.05-0.10 mm (0.002-0.004 inch) is the correct thickness of shims (90) to be removed to obtain correct preload.

To assemble drive shaft assembly (8 through 27), clamp drive shaft (19) in a protected vise so that internal splined bore is downward. Install 2-6 gear (18) and synchronizer ring (12) on shaft. Slide shift sleeve (13) on shaft with flat side facing downward. Select and install snap ring (11) of correct thickness to provide shift sleeve end play of 0-0.1 mm (0-0.004 inch). Snap rings (8 and 11) are available in thickness of 1.45, 1.55, 1.65 and 1.75 mm (0.057, 0.061, 0.065 and 0.069 inch). Install springs (16), ball pins (15) and thrust blocks (14), then slide shift collar into place so that third tooth beside each tooth gap is in mesh with corresponding thrust block (Arrow—Fig. 129). Install second synchronizer ring (12—Fig. 124), 1-5 gear (10) and thrust washer (9). Select and install snap ring (8) of correct thickness to provide 1-5 gear (10) end play of 0.0-0.4 mm (0.0-0.016 inch). Invert drive shaft (19) in vise and slide 3-7 gear (23) into position on drive shaft. Install second synchronizer in same manner as the first one. Make certain that lube groove in shift sleeve (25) is aligned with lube bores in drive shaft. Select and install snap ring (27) of correct thickness to provide shift sleeve end play of 0-0.1 mm (0-0.004 inch). Snap rings (27) are available in thicknesses of 1.55, 1.65 and 1.75 mm (0.061, 0.065 and 0.069 inch).

Adjust cone point of bevel pinion shaft (44) by installing correct shim pack (47) between bearing cup (46) and transmission case. Measure height of bearing cup (46) and cone (45). Add this measurement to the dimension etched in mm on end of bevel pinion shaft. Subtract the sum of these dimensions from the dimension stamped in identification plate on right side of transmission case. The remainder will be the thickness of shim pack (47) to be installed.

To adjust the bearing preload on bevel pinion shaft (44), install bevel pinion shaft and countershaft cluster gear assembly (44 through 58) in transmission case. Tighten special nut (58 or 58A) to a torque of 140 N·m (100 ft.-lbs.). Engage shift collar (49) in range II. Wrap a string around cluster gear and attach a spring scale to the string to check rolling torque. Rolling drag torque should be 0.75-1.50 N·m (6.5-13 in.-lbs.) with new bearings or 0.40-0.75 N·m (3.5-6.5 in.-lbs.) with used bearings; but when checked with string as described, pull should be 20-40 N (4.5-9 lbs.) with new bearings, 10-20 N (2.2-4.5 lbs.) with used bearings. If necessary, add or remove shims (55) to obtain correct rolling drag torque.

Adjust bearing preload on extension shaft as follows: On models without front-wheel drive, install extension shaft (59 or 59A) on bevel pinion shaft splines. On later models, lugs on extension shaft must engage groove of the later special nut (58A). Install bearing cup (69) so it protrudes about 2 mm (0.078 inch) out front face of bearing quill (70). Install low range I-reverse shift fork in shift collar (63). Attach bearing quill (70) and tighten retaining cap screws to a torque of 50 N·m (35 ft.-lbs.). Install a thickness of shims (71) between quill and plate (72) sufficient to obtain a measurable end play with cap screws (73) tightened. Measure extension shaft end play. This measurement plus 0.05-0.10 mm (0.002-0.004 inch) is the correct thickness of shims (71) to be removed to obtain correct preload.

On models equipped with front-wheel drive, install extension shaft (74 or 74A) with gears and shift mechanism on bevel pinion shaft splines. On later models, lugs on extension shaft must engage groove of the later special nut (58A). Press bearing cup (76) into bearing quill (70) without shims (78). Attach bearing quill to transmission case and tighten retaining cap screws to a torque of 50 N·m (35 ft.-lbs.). Measure extension shaft end play. This measurement plus 0.05-0.10 mm (0.002-0.004 inch) is the thickness of shim pack (78) to be installed for correct preload.

To adjust preload of bearings on top shaft assembly (Fig. 130), install top shaft assembly and with bearing cups (2 and 32—Fig. 124) in place, but do not install snap ring (1). Install snap ring (33) or retainer plate (33A). If retainer plate (33A) is used, coat threads of Torx screws with Loctite 242 and tighten to 60 N·m (45 ft.-lbs.) torque. Push assembly, including bearing cup (2), toward front and install the thickest snap ring (1) that can be installed in groove. Install gear (30B) if so equipped, install bearing cup (36) in bearing quill; then install bearing quill. Tighten bearing quill cap screws to a torque of 50 N·m (35 ft.-lbs.). Slide spacer (34) with chamfered side rearward on shaft (30). Install adequate shims (35) to provide measurable shaft end play, with bearing cone (37) and special nut (38) installed. Tighten the special nut (38) to 140 N·m (100 ft.-lbs.) torque and measure shaft end play. This measurement plus 0.05-0.10 mm (0.002-0.004 inch) is the correct thickness of shims (35) to be removed to obtain correct bearing preload. Install bearing cap (B—Fig. 128) and tighten cap screws to 120 N·m (85 ft.-lbs.) torque. Rolling torque of the assembled top shaft should be 1-2 N·m (9-18 in.-lbs.) and can be checked using a torque wrench on nut (38—Fig. 124).

Refer to paragraph 123 and complete the reassembly of tractor.

Fig. 129—When installing shift collar, align third tooth of shift collar with thrust block (14) as shown at arrow.

Fig. 130—On later tractors, align arrow mark on shaft (59A or 74A—Fig. 124) with groove in special nut (58A). If shaft will not engage groove in nut, turn shaft slightly and re-try.

TRANSMISSION OIL PUMP

All Models with Synchronized Transmission

The transmission oil pump used is an internal gear, crescent-type pump located in the clutch housing and driven by the pto clutch drive shaft. The pump supplies oil for the Hi-Lo clutch, pto clutch, front-wheel drive clutch and the main hydraulic pump. Oil from the transmission oil pump is also delivered to the oil cooler, to tractor brakes and to the lubricating systems for the Hi-Lo unit and synchronized transmission.

127. A two stage bypass valve (1 through 6—Fig. 131) is located above the filter in transmission housing. If the filter (16) becomes clogged or if oil is too cold to circulate through filter easily, the differential pressure (before the filter and after the filter) will exceed the opening pressure of the bypass valve. At the first stage (about 500 kPa or 70 psi difference), unfiltered oil will be directed into the lube circuits. When the pressure differential exceeds 650 kPa (90 psi), unfiltered oil is still directed to the lubrication circuits, but a larger volume of oil is directed back to the main oil reservoir (transmission case). Be sure that shoulder on valve (3) is down when installing. Specifications for spring (5) should be as follows:

Free length . . . . Approx. 68 mm
                                                 (2.68 in.)
Pressure when compressed to 49 mm . . . . . . 50-60 N
            (1.5 in.)   (11-13 lbs.)
Pressure when compressed to 38 mm . . . . . . 80-95 N
            (1.93 in.)   (18-21 lbs.)

If low flow volume is suspected, remove, clean and inspect transmission pump inlet screen (7). If screen is clean, separate tractor between clutch housing and transmission case and check the following:

  a. Pump inlet and discharge tube "O" rings and packings for deterioration.
  b. Pump body screws for proper torque.
  c. Pump body and gears for excessive wear or scoring.

128. R&R AND OVERHAUL. To remove the transmission oil pump, first split tractor between clutch housing and transmission as outlined in paragraph 123. If so equipped, remove Hi-Lo unit, Creeper transmission, front-wheel drive clutch and pto clutch. Remove oil pump suction and pressure lines and retainers. Remove the three oil pump retaining cap screws, then pull oil pump with pto clutch drive shaft out of clutch housing.

Remove cap screws (6—Fig. 132) and separate pto clutch drive shaft (4) from pump assembly. Remove snap ring (9) and press bearing (8) from shaft. Remove snap ring (1), then using a slide hammer puller, remove needle bearing (2), oil seal (3) and cap (3A). Remove cap screws (10) and separate bearing housing (12) from pump housing and carrier (17). Remove spur gear (15) and internal gear (14) from pump housing and carrier. Remove oil seal (16) and "O" ring (18).

Clean and inspect all parts and renew any showing excessive wear or other damage. Bearing housing (12), internal gear (14), spur gear (15) and pump

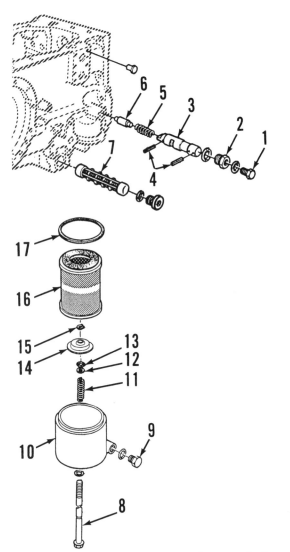

Fig. 131—View showing locations of transmission oil pump screen (7), filter (16) and bypass valve (3).

1. Plug
2. Threaded nipple
3. Bypass valve
4. Spring pins
5. Spring
6. Pin
7. Inlet screen
8. Bolt
9. Plug
10. Filter cover
11. Compression spring
12. Washer
13. "O" ring
14. Retainer
15. Snap ring
16. Filter
17. Seal washer

housing and carrier (17) are available only as an assembly.

Reassemble oil pump by reversing disassembly procedure, keeping the following points in mind: Install gears (14 and 15) in pump housing and carrier (17), chamfered side first. Place a straightedge across sealing surface of housing and carrier and measure clearance between gears and straightedge. Clearance should be 0.03-0.05 mm (0.001-0.002 inch). Clean sealing surfaces of pump housing and carrier (17) and bearing housing (12) with Loctite primer, then coat with Loctite 515. Tighten cap screws (6 and 10) to a torque of 55 N·m (40 ft.-lbs.). Fill pump with transmission oil and check pump for free rotation. Coat sealing surface of clutch housing partition with Loctite 515 and coat carrier sleeve with Moly High Temperature EP grease. Install oil pump assembly and tighten cap screws to a torque of 55 N·m (40 ft.-lbs.).

The balance of reassembly is the reverse of disassembly. Refer to paragraph 123 and rejoin tractor.

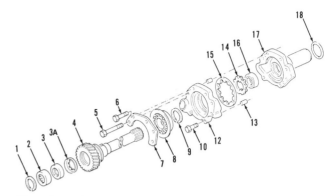

Fig. 132—Exploded view of transmission oil pump assembly typical of all models with synchronized transmission.

1.	Snap ring	9.	Snap ring
2.	Needle bearing	10.	Cap screw (2)
3.	Oil seal	12.	Bearing housing
3A.	Cap (late models)	13.	Dowel pins (2)
4.	Pto clutch drive shaft	14.	Internal gear
5.	Cap screw (3)	15.	Spur gear
6.	Cap screw (2)	16.	Oil seal
7.	Retainer	17.	Pump housing & carrier
8.	Ball bearing	18.	"O" ring

DIFFERENTIAL AND FINAL DRIVES

The differential is a four-pinion-type and is equipped with a differential lock. See Fig. 133 and Fig. 135.

Final drives incorporate a planetary gear reduction unit at inner end of the housing.

DIFFERENTIAL

All Models

129. REMOVE AND REINSTALL. To remove differential, drain transmission and remove final drives as outlined in paragraph 135 and rockshaft housing as in paragraph 166.

The differential lock must be removed as follows: Remove clamp screw and lever (23—Fig. 133) from models without cab, or remove bellcrank from end of shaft (20) on models with Sound Gard Body; then remove the square key (21) from all models. Hold fork (16) in place, bump shaft (20) rearward and remove key (19) and plug (17). Remove shaft, fork and shift collar (1).

Remove the transmission oil cup and rear oil lines. Remove the rockshaft control arm and cam from transmission case. Support differential, unbolt both bearing quills (2 and 11), remove quills, then lift out differential. Be sure to keep shims (3) with correct quill to aid in reassembly. Shims are located between quills and transmission case to adjust backlash of the main drive bevel gears and to adjust preload of differential carrier bearings.

When reassembling, first check and adjust mesh position as described in paragraph 132, then check and adjust bearing preload as in paragraph 133 and backlash as in paragraph 134.

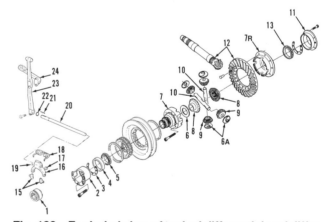

Fig. 133—Exploded view of typical differential and differential lock assembly. Some models do not use thrust washers (6 and 6A). Bevel pinion shaft and ring gear (12) are available only as a matched set for all models.

1.	Differential lock collar	12.	Bevel pinion shaft
2.	Left quill		& ring gear
3.	Shims	13.	Bearing cone
4.	Bearing cup	15.	Shoes
5.	Bearing cone	16.	Fork
6.	Thrust washer	17.	Plug
6A.	Thrust washer	18.	Spring
7.	Left housing half	19.	Woodruff key
7R.	Right housing half	20.	Shaft
8.	Side gears	21.	Square key
9.	Pinions	22.	"O" ring
10.	Pinion shafts	23.	Lever
11.	Right quill	24.	Pedal

130. OVERHAUL. The bevel ring gear, bevel pinion shaft and the right half of the differential housing are available only as a matched set.

To disassemble the removed differential assembly, remove the eight cap screws that attach the differential housing halves together. Procedure for removing bearing cup (4—Fig. 133) and bearing cones (5 and 13) will be obvious. Cup for bearing cone (13) is integral with quill (11). If pinions (9) or shafts (10) are worn or damaged, all mating parts should be renewed. Check condition of differential housing bores if side gears (8) are worn.

Thrust washer (6) . . . . 1.95-2.05 mm
                                             (0.077-0.081 in.)
Thrust washer (6A) . . . . 0.95-1.05 mm
                                             (0.037-0.041 in.)

Reassemble by reversing disassembly procedure. Coat shaft (20) with Molykote grease before assembling. Tighten cap screws attaching differential housing halves together, bearing quill to transmission case cap screws and the differential lock operating lever clamp screw all to 50 N•m (35 ft.-lbs.) torque. Tighten screws attaching final drives to transmission housing to 230 N•m (170 ft.-lbs.) for 2955 models; 120 N•m (85 ft.-lbs.) for all other models. Rockshaft housing to transmission case cap screws should be tightened to 120 N•m (85 ft.-lbs.) torque.

MAIN DRIVE BEVEL GEARS

All Models

131. ADJUSTMENT. If differential is removed for access to other parts and no defects in the adjustments are noted, the shim pack (53—Fig. 116) and (3—Fig. 133) should be kept intact and reinstalled in their original positions. However, if bevel gears, bearings, quills or transmission case are renewed, the main bevel gears should be checked for mesh (cone point) position, differential carrier bearing preload and gear backlash.

132. MESH (CONE POINT) POSITION. The fore and aft position of the bevel pinion shaft is controlled by shims located between pinion shaft rear bearing cup and front wall of differential compartment. Shims for collar shift transmission is shown at (39—Fig. 116) and at (47—Fig. 124) for synchronized transmission. When renewing parts, the shim pack required for correct mesh position of the bevel pinion shaft and ring gear can be determined as follows:

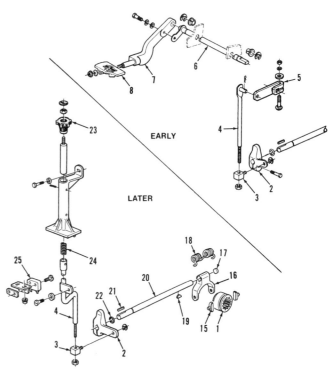

Fig. 135—Exploded view of differential lock linkage used on models equipped with Sound Gard Body.

1.	Differential lock collar		
2.	Lever	17.	Plug
3.	Swivel	18.	Spring
4.	Rod	19.	Woodruff key
5.	Arm	20.	Shaft
6.	Shaft	21.	Square key
7.	Lever	22.	"O" ring
8.	Pedal	23.	Handle
15.	Shoes	24.	Spring
16.	Fork	25.	Switch

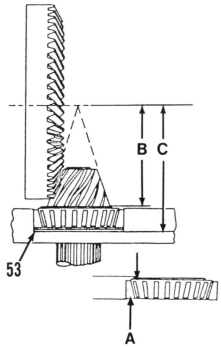

Fig. 134—Refer to text for setting mesh position of the main drive bevel gears.

Measure height of the rear pinion bearing cup and cone (A—Fig. 134). Add 0.10 mm (0.004 inch) to this measurement to compensate for change in height caused by pressing cup into housing. Add this measurement to the dimension etched in mm on end of bevel pinion shaft. Subtract the sum of these dimensions from the dimension stamped on rear top of transmission case or on data plate attached to transmission case. The remainder will be the thickness of shim pack (S) to be installed. Refer to paragraph 119 or paragraph 126 for adjustment of bevel pinion shaft bearing preload.

133. DIFFERENTIAL BEARING ADJUSTMENT. Differential carrier bearings should be adjusted to a preload of 0.08-0.13 mm (0.002-0.005 inch) as follows: Install differential and bearing quills with original shim packs, then check differential end play using a dial indicator.

> NOTE: When making this adjustment, be sure that clearance exists between the main drive bevel ring gear and pinion shaft at all times.

If no differential end play exists, add shims under right bearing quill to obtain not more than 0.05 mm (0.002 inch) end play. If more than 0.05 mm (0.002 inch) end play existed on original check, subtract shims.

On 2750 and 2955 models, measure end play of differential, then subtract shims equal to the measured end play plus an additional 0.15-0.25 mm (0.006-0.010 inch) to give the desired bearing preload of 0.15-0.25 mm (0.006-0.010 inch).

On 2755 and 2855N models, measure end play of differential, then subtract shims equal to the measured end play plus an additional 0.08-0.13 mm (0.003-0.005 inch) to give the desired bearing preload of 0.05-0.13 mm (0.002-0.005 inch).

Shims are available in thicknesses of 0.08, 0.13 and 0.25 mm (0.003, 0.005 and 0.010 inch).

134. BACKLASH ADJUSTMENT. With differential carrier bearing preload adjusted as in paragraph 133, adjust backlash between bevel pinion shaft and ring gear to 0.3 mm (0.012 inch) by transferring bearing quill shims (3—Fig. 133) from one side to the other as required. Moving shims from right to left will decrease backlash. Do not change total thickness of all shims during backlash adjustment or the previously determined preload adjustment will be changed. Set backlash as close to 0.3 mm (0.012 inch) as possible. Measure backlash at several locations around ring gear. Backlash should not be less than 0.22 mm (0.009 inch), nor more than 0.38 mm (0.015 inch) at any position around ring gear. Improper installation of ring gear is indicated if these limits are exceeded.

FINAL DRIVES

All Models

135. REMOVE AND REINSTALL. To remove a final drive unit, disconnect battery ground cables and drain transmission. Support rear of tractor and remove wheel and tire assembly. Remove rear fenders and roll guard. If right final drive is being removed and tractor has selective control valve, disconnect pressure line, coupler lines and return hose between valve and rockshaft housing and remove control valve. Disconnect brake line from final drive housing. Attach hoist to final drive, remove attaching cap screws and pull final drive assembly from transmission case. If tractor is equipped with Sound Gard Body and only one final drive is to be removed, it may be possible to disconnect body and raise enough to permit removal; however, if both final drives are to be removed, the body should be removed. Refer to paragraph 174 for removing and installing Sound Gard Body.

Fig. 136—Shaft (20) can be removed from lever (2) using a prying tool (5) to push the lever and shaft (20) toward rear, loosen the lever clamping screw, then move the lever forward on shaft. Tighten clamp on lever and pry the lever and shaft toward rear again until lever can be removed.

Fig. 137—Use a screwdriver and snap ring pliers to remove snap ring (28).

Reinstall by reversing removal procedure and tighten screws attaching final drives to transmission housing to 230 N·m (170 ft.-lbs.) torque for 2955 models; 120 N·m (85 ft.-lbs.) torque for all other models.

136. OVERHAUL. To overhaul the removed final drive unit, first remove brake housing with ring gear (10—Fig. 138) from final drive housing (19). Remove lock plate (25), unscrew special twelve-point cap screw (26), then pull planetary carrier assembly from axle. Drive rear axle shaft out of final drive housing and remove bearing cone (5). Remove snap ring (4), bearing cone (22) and oil seal (1). To remove planet gears (33), expand snap ring (28), lift it from groove in carrier (29) and pull planet shafts (34) out far enough so snap ring can be removed. See Fig. 137. Remove planet shafts, planet gears and bearing nee-

dles. Check all planetary parts for pitting, scoring or excessive wear and renew parts as required. If any planet gear bearing needles are defective, renew the complete set.

Reassemble final drive and adjust axle bearings as follows: Coat bores of planet gears with grease and place needles in gears. Install spacer (32—Fig. 138) between the two rows of bearing needles and position a thrust washer at each side. Place planet gears in carrier and insert planet shafts (34) only far enough to retain bearing needles and thrust washers. Install snap ring (28—Fig. 137) in slots of planet shafts, then complete insertion of shafts and be sure snap ring seats in groove in carrier.

Heat bearing cone (22—Fig. 138) to 150° C (300° F) and drive on rear axle shaft. Coat inner seal (7) with grease and install axle in housing. Heat bearing cone (5) to not more than 150° C (300° F), and install on inner end of axle shaft. Place carrier assembly on axle, install retaining washer (27) and cap screw (26). Tighten cap screw until bearing is pulled into place and a small amount of axle end play remains. Check the amount of torque required to turn the axle with the existing axle end play. Then, tighten the cap screw to increase the rolling torque an additional 10-14.5 N·m (90-126 in.-lbs.) for 2750, 2755 and 2855N models; 10-13.5 N·m (90-120 in.-lbs.) for 2955 models. Install lock plate (25). Use new gaskets (9 and 13) and reinstall brake housing and final drive. Lubricate outer axle bearing (22) with 6-8 strokes of EP multi-purpose grease. Refer to Fig. 139 for assembled view.

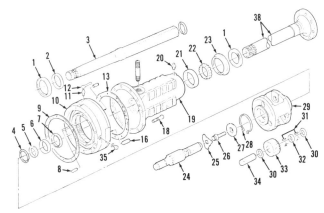

Fig. 138—Exploded view of final drive assembly.

1. Oil seal		20. Grease fitting
2. Spacer		21. Bearing cup
3. Axle shaft		22. Bearing cone
4. Snap ring		23. Seal cup
5. Bearing cone		24. Final drive shaft
6. Bearing cup		25. Lock plate
7. Oil seal		26. Special cap screw
8. Dowel pin		27. Washer
9. Gasket		28. Snap ring
10. Brake housing with		29. Planet carrier
ring gear		30. Washers
11. Brake bleed valve		31. Bearing needles
12. Cap		32. Spacer
13. Gasket		33. Planet gear (3)
16. Dowel pin		34. Planet shaft (3)
18. Cap screw		35. Plug
19. Final drive housing		38. Flanged axle shaft

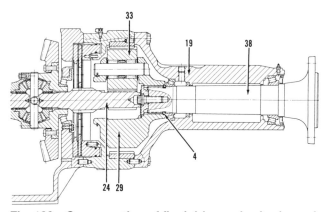

Fig. 139—Cross section of final drive and axle shown in Fig. 138.

BRAKES

The brakes are hydraulically actuated and use a wet-type disc controlled by a brake operating valve located on right side of clutch housing on models not equipped with Sound Gard Body, or on right side of dash panel support on models with Sound Gard Body. Brake discs are splined to the final drive shafts and the brake pressure ring is fitted in inner end of final drive housing. Except for a pedal adjustment, no other brake adjustments are required.

BLEED AND ADJUST

All Models

137. BLEEDING. Brakes must be bled when pedals feel spongy, pedals bottom, or after disconnecting or disassembling any portion of the brake system.

To bleed brakes, start engine and operate for at least two minutes at 2100 rpm, turning steering wheel from left to right and back to left to ensure that air is removed from steering system and that brake control valve reservoir is filled. Attach a clear plastic bleed hose to brake bleed screw (11—Fig. 138) located on top side of final drive housing and place opposite end in filler hole of rockshaft housing. Loosen bleed screw about ¾ turn. Depress brake pedal being bled. Before pedal reaches the end of its travel, close bleeder screw and let pedal return. Continue sequence until oil in bleed hose is free of air bubbles.

NOTE: If engine is not running, engine must be started and brake valve reservoir refilled after each 15 strokes of the pedal.

Remove bleed hose and repeat operation on the other brake.

138. PEDAL ADJUSTMENT. Before making this adjustment, bleed brakes as in paragraph 137. For access to brake control valve on models with Sound Gard Body, raise platform mat, then remove hand brake lever cover, hand brake lever, rear dash panel and right side dash panel.

On all models, adjust stop screws (1 and 2—Fig. 140) so that brake pistons are fully extended from housing and brake pedal arms are just touching the pistons. Then, turn both stop screws out an additional ½ turn on tractors before serial number 623 079L; ⅚ turn for tractors after serial number 623 078L. If pedals are not the same height, turn the stop screw on highest pedal out not more than ⅙ turn until pedals are even.

BRAKE TEST

All Models

139. PEDAL LEAK-DOWN. With a 270 N (60 lbs.) pressure applied continuously to each pedal for one minute, the pedal leak-down should not exceed 25 mm (1 inch). Excessive brake pedal leak-down can be caused by air in the brake system, faulty brake control valve pistons and/or "O" rings, faulty brake pressure ring seals, or faulty brake control valve equalizing valves or reservoir check valves. Faulty brake control valve pistons or "O" rings will be indicated by external leakage around the pistons.

Faulty pressure ring seals, or brake control valve, can be determined as follows: Isolate brake from brake valve by plugging brake line. If leak-down stops, pressure ring seals are defective. If leak-down continues, brake control valve is faulty and can be checked further by depressing brake pedals individually, then simultaneously. If leak-down occurs in both cases, a defective reservoir check valve is indicated. If leak-down occurs during individual pedal operation, but not on simultaneous pedal operation, a faulty equalizer valve is indicated.

Refer to appropriate following paragraphs for brake control valve and brake pressure ring overhaul procedures.

R&R AND OVERHAUL

All Models

140. BRAKE CONTROL VALVE. To remove the brake control valve from models not equipped with Sound Gard Body, thoroughly clean valve and surrounding area. Disconnect brake lines from rear of control valve and cap or plug openings. Remove retaining ring (21—Fig. 141) and remove pedal shaft

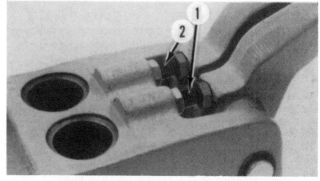

Fig. 140—View of brake pedal adjusting screws on models not equipped with Sound Gard Body. Adjustment is similar for models with Sound Gard Body.

(20). Remove brake pedals, then unbolt and remove valve assembly.

On models equipped with Sound Gard Body, raise platform mat, then remove hand brake lever cover, hand brake lever, rear dash panel and right side dash panel. Thoroughly clean brake valve and surrounding area. Disconnect oil lines from inlet elbow (30—Fig. 142) and outlet elbow (34), then disconnect brake lines from elbow fittings (1). Cap or plug all line openings. Unbolt and remove brake valve assembly. Remove retaining ring (21), withdraw shaft (20) and remove brake pedals (29).

On all models, remove fittings (1—Fig. 141 or Fig. 142), springs (5) and balls (6). Remove seats (3) and ball retainers (7), then push pistons (9) and springs (8) out of valve bores. Remove expansion plugs (11) or connectors (32) with inlet and outlet elbows (30 and 34), then remove reservoir check valve assemblies (items 13, 14, 15 and 16). Remove equalizer valve assemblies (items 22, 23, 24 and 25). "O" rings (18) and oil seals (19) can be removed from piston bores.

Clean and inspect parts for wear, scoring, cracks or other damage and renew as required. Check springs as follows:

Brake piston springs (8)—
Free length . . . . 190 mm
                                  (7.5 in.)
Test pressure at 145 mm . . . . 90 N
       (5.75 in.)   (20 lbs.)

Return oil regulating springs (5)—
Free length . . . . 11 mm
                                    (0.43 in.)
Test pressure at 6 mm . . . . 0.8-0.9 N
       (0.24 in.)   (2.7-3.3 oz.)

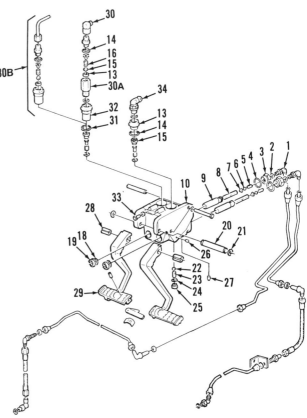

Fig. 142—Exploded view of brake control (master cylinder) valve typical of models with Sound Gard Body. Parts shown with adapter (30A) are used on early models; parts (30B) are used on later models.

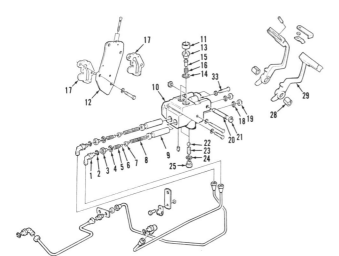

Fig. 141—Exploded view of brake control (master cylinder) valve typical of models without Sound Gard Body.

1. Fitting	15. Valve	
2. "O" ring	16. Spring	
3. Check valve seat	17. Gasket	
4. "O" ring	18. "O" ring	
5. Spring	19. Oil seal	
6. Steel ball	20. Pedal shaft	
7. Retainer	21. Retaining ring	
8. Spring	22. Steel ball	
9. Piston	23. Spring	
10. Valve housing	24. "O" ring	
11. Expansion plug	25. Plug	
12. Intermediate plate	28. Pedal bushing	
13. Check valve seat	29. Pedals	
14. "O" ring	33. Adjusting screw	

1. Elbow fitting		
2. "O" ring		21. Retaining rings
3. Check valve seat		22. Steel ball
4. "O" ring		23. Spring
5. Spring		24. "O" ring
6. Steel ball		25. Plug
7. Retainer		26. Set screw
8. Spring		27. Plug
9. Piston		28. Pedal bushing
10. Valve housing		29. Brake pedals
13. Check valve seat		30. Oil inlet elbow
14. "O" ring		30A. Early type adapter
15. Check valve		30B. Later type
16. Spring		31. "O" ring
18. "O" ring		32. Connector
19. Oil seal		33. Adjusting screw
20. Brake pedal shaft		34. Oil return elbow

Check valve springs (16)—
  Free length . . . . 23 mm
                                         (0.90 in.)
  Test pressure at 8 mm . . . . 0.4-0.6 N
        (0.32 in.)   (1.4-2.0 oz.)

Equalizer valve springs (23)—
  Free length . . . . 20 mm
                                         (0.80 in.)
  Test pressure at 7 mm . . . . 0.6-0.8 N
        (0.28 in.)   (2.0-2.7 oz.)

Renew housing if seats for equalizer balls (22) are damaged. Oil seals (19) are installed with lips toward outside. Pay particular attention to area of reservoir check valves (15) where contact is made with brake pistons and renew valves if any doubt exists as to its condition. Renew bushings and/or pedal shaft (20) if clearance is excessive. Lubricate all parts when reassembling.

Use new gasket (17—Fig. 141) when installing brake valve on tractor not equipped with Sound Gard Body. Reinstall brake valve by reversing removal

Fig. 143—Brake disc is splined to final drive shaft. Model shown is typical.

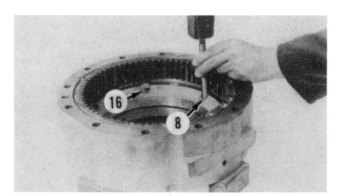

Fig. 143A—Brake pressure ring can be driven from bore in housing driving a punch evenly against the three dowel pins (8). Be sure that reaction pins (16) do not bind.

procedure. Bleed brakes as in paragraph 137 and adjust pedals as in paragraph 138.

141. BRAKE PRESSURE RING, PLATE AND DISC. To remove pressure ring, plate and disc, remove final drive as outlined in paragraph 135. Pull final drive shaft from differential and remove brake disc from final drive shaft. See Fig. 143. Lift brake pressure plate from dowels in transmission case. Carefully remove pressure ring (Fig. 143A) from brake housing.

Inspect brake disc (2—Fig. 144) for worn or damaged facings or damaged splines and renew as necessary. Inspect pressure plate for scoring, checking or other damage and renew as required.

To reassemble, install new seals on pressure ring (3), lubricate and install in brake housing over dowel pins. Make sure that neither seal is cut or rolled during installation. Place pressure plate over dowels in transmission case. Install brake disc on final drive shaft so thickest facing is next to transmission case and insert shaft into differential. Install final drive housing.

Bleed brakes as outlined in paragraph 137.

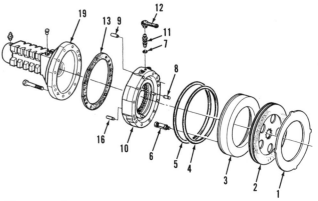

Fig. 144—Exploded view of typical brake assembly. Differences may be noted.

1. Actuating disc	9. Dowel pins (3 used)
2. Brake disc	10. Housing & ring gear
3. Brake pressure ring	11. Bleeder screw
4. "O" ring	12. Cap
5. "O" ring	13. Gasket
6. Reaction pin & spring assy.	16. Dowel pin
7. "O" ring	19. Axle housing

PARKING BRAKE

R&R AND OVERHAUL

2750, 2755 and 2855N Models

142. On models with Sound Gard Body, remove the body as outlined in paragraph 174. On all tractors, drain transmission, remove rockshaft housing assembly as in paragraph 166 and remove the left final drive assembly as outlined in paragraph 135. If the brake drum (9—Fig. 145 or Fig. 145A) is to be removed, refer to paragraph 129 and remove the differential.

On models without body, remove hand brake lever (1—Fig. 145) and Woodruff key from shaft (22) or lever (21). On models with Sound Gard Body, remove lever (27—Fig. 145A) and Woodruff key from shaft (28). On all models, remove support screw (11—Fig. 145 or Fig. 145A) from bottom of transmission case and remove brake adjusting screw (15) from threaded end of brake band (12). Use a 3/8 inch-16 UNC cap screw to pull upper anchor pin (19) and lower anchor pin (20) from bores. Remove pin and detach link (18) from shaft (22), then remove shaft (22). Pull anchor (17) up far enough to remove pin (13), then pull

anchor (17) and associated parts from transmission case. Rotate brake band enough to remove spring (14), unhook spring (14), then remove band (12).

Clean and inspect all parts for excessive wear or other damage and renew as required. If new linings are riveted to band, heads should be toward outside. Reinstall hand brake by reversing the removal procedure and coat brake shaft (22) and bore in case with Moly High Temperature EP grease. If new hand brake linings (12) have been installed, adjust band support screw (11) by loosening two turns and tightening lock nut. Readjust support screw after about 100 hours to the standard setting.

Adjust brake as follows: Apply hand brake lever so that latch (5) engages first notch of quadrant (6). Tighten brake band adjusting screw (15) by hand using a screwdriver. Tighten band support screw (11) at bottom of transmission case by hand using a screwdriver, then back off 1/2 turn and secure with locknut. Adjust brake band adjusting screw (15) until hand brake lever latch (5) will engage in third or fourth notch in quadrant when lever is pulled with a force of about 110 N (25 lbs.). Refer to Fig. 145B for assembled view of hand brake linkage.

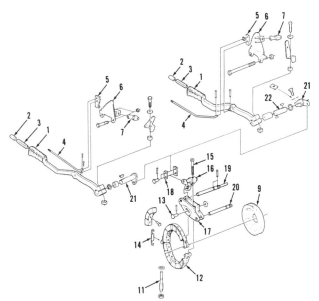

Fig. 145—Exploded view of hand brake assembly typical of 2750, 2755 and 2855N models without Sound Gard Body.

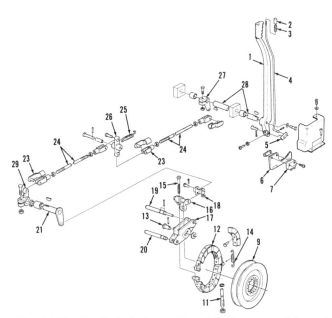

Fig. 145A—Exploded view of hand brake assembly typical of 2750, 2755 and 2855N models with Sound Gard Body. Refer to Fig. 145 for legend except for the following.

1. Brake lever
2. Release button
3. Spring
4. Release rod
5. Latch
6. Quadrant
7. Brake light switch
9. Brake drum
11. Band support screw
12. Brake band
13. Pin
14. Spring
15. Adjusting screw
16. Brake band arm
17. Anchor
18. Links
19. Upper anchor arm pin
20. Lower anchor arm pin
21. Lever
22. Brake shaft

23. Yoke
24. Brake rod
25. Spring
26. Pivot arm
27. Lever
28. Shaft
29. Arm

2955 Models

143. On models with Sound Gard Body, remove the body as outlined in paragraph 174. On all tractors, drain transmission and remove rockshaft housing assembly as in paragraph 166. Remove brake outer lever, then remove clamp screw from internal lever. Drive the operating shaft inward, remove Woodruff key from shaft, then remove the shaft. Turn adjuster (15—Fig. 147 or Fig. 147A) from threaded bolt of lower brake band (12L), remove brake band lower support screw (11) from bottom of transmission case, then remove nuts (34) attaching top section of brake band to the center section. Remove band retaining pin (36) and lift top section of band (12U) from tractor. Move bottom section of band (12L) down, then remove band arm (16) with links (18) and connector (30) toward rear. Turn the remaining brake band sections around drum (9) until center section (12C) can be unbolted from the lower section (12L), then remove lower and center sections of brake band.

To remove brake drum and hub (9 and 31), first remove differential as outlined in paragraph 129.

Clean and inspect all parts for excessive wear or other damage and renew as required. If new linings are riveted to band, heads should be toward outside. Reinstall hand brake by reversing the removal proce-

dure and coat brake shaft (22) and bore in case with Moly High Temperature EP grease. If new hand brake linings (12) have been installed, adjust band support screw (11) by loosening two turns and tightening lock nut. Readjust support screw after about 100 hours to the standard setting.

Adjust brake as follows: Apply hand brake lever so that latch (5) engages first notch of quadrant (6). Tighten brake band adjusting screw (15) by hand using a screwdriver. Tighten band support screw (11) at bottom of transmission case by hand using a screwdriver, then back off ½ turn and secure with locknut. Adjust brake band adjusting screw (15) until hand brake lever latch (5) will engage in third or fourth notch in quadrant when lever is pulled with a force of about 110 N (25 lbs.). Refer to Fig. 147B for assembled view of hand brake linkage.

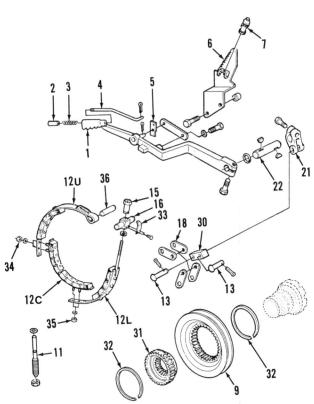

Fig. 147—Exploded view of hand brake assembly typical of 2955 models without Sound Gard or Roll Gard.

1.	Brake lever		
2.	Release button		
3.	Spring	13.	Pin
4.	Release rod	15.	Adjusting screw
5.	Latch	16.	Brake band arm
6.	Quadrant	18.	Links
7.	Brake light switch	21.	Lever
9.	Brake drum	22.	Brake shaft
11.	Band support screw	30.	Connector block
12C.	Brake band	31.	Brake hub
	(center section)	32.	Snap rings
12L.	Brake band	33.	Leaf spring
	(lower section)	34.	Nut
12U.	Brake band	35.	Nut
	(upper section)	36.	Anchor pin

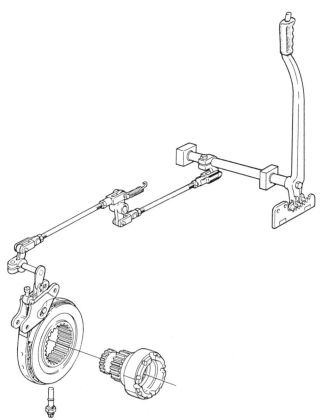

Fig. 145B—Drawing of hand brake assembly typical of 2750, 2755 and 2855N models with Sound Gard body.

Illustrations for Fig. 145B and Fig. 147 reproduced by permission of Deere & Company. Copyright Deere & Company.

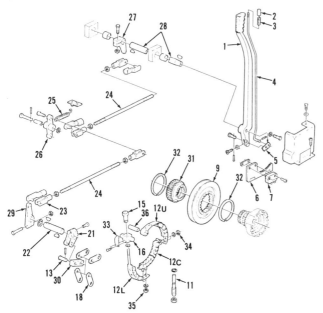

Fig. 147A—Exploded view of hand brake assembly typical of 2955 models with Sound Gard or Roll Gard. Refer to Fig. 147 for legend except the following.

23. Yoke
24. Brake rod
25. Spring
26. Pivot arm
27. Lever
28. Shaft
29. Arm

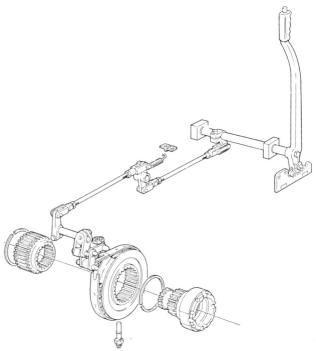

Fig. 147B—Drawing of hand brake assembly typical of 2955 models with Sound Gard body.

POWER TAKE-OFF (INDEPENDENT)

OPERATION

All Models So Equipped

144. The Independent Power Take-Off may provide only one speed (540 rpm) or dual speeds (either 540 rpm or 1000 rpm) operation. On dual speed pto, the speed will depend upon which stub shaft (20 or 21—Fig. 153) is installed. On all models, engagement of the multiple disc hydraulic clutch or the pto brake is accomplished by directing oil with the pto control valve (Fig. 149). The pto control valve is designed so there is no overlap between clutch and brake circuits. When control lever is moved to the engaged position, pressure fluid flows to the pto clutch and at the same time enters the area behind the valve. This prevents rapid opening of the valve and consequent rough clutch engagement. At the same time, pressure is ported to the clutch, pto brake pressure is released and the brake piston passage is opened to the sump.

When control lever is moved to the disengaged position, clutch passage is opened to the sump and pressure is applied to a piston located in countershaft bearing support that applies a disc-type brake to pto clutch shaft.

> NOTE: When there is a drop in hydraulic pressure or when engine is shut off, the hydraulic shift lever arm detent releases and control lever automatically shifts to disengaged (brake) position.

PRESSURE TEST

All Models So Equipped

145. On tractors without Sound Gard Body, remove shift lever shield from transmission shift cover. On models equipped with Sound Gard Body, remove floor mat and floor plate. On all models, to check pto brake pressure, remove plug at (E—Fig. 148) and install test gage. To check pto clutch pressure, remove plug (A) and install test gage. With pto control lever in neutral position, start engine and operate at 1500 rpm. Move lever to brake or clutch position. Gage should read 1000 kPa (145 psi) in either position.

With gages installed in ports (A and E) and engine operating at 1500 rpm, observe both gages, Slowly move control lever from engaged to disengaged position. One gage must read zero before other gage shows any pressure. No pressure overlap should be experienced when shifting slowly between engaged and disengaged positions.

R&R AND OVERHAUL

All Models So Equipped

146. CONTROL VALVE. The control valve is located in the transmission shift cover. On models without Sound Gard Body, remove shift lever shield from transmission shift cover. On models equipped with Sound Gard Body, remove floor mat and floor plate. On all models, disconnect pto shift linkage from control lever shaft (16 or 18—Fig. 149). Remove cover (20) with shift arm detents (25 and 30) and control lever shaft. Remove transmission shift cover assembly (32). Remove all parts of detents (25 and 30) from cover (20). Drive plug (33) out of transmission shift cover, then remove pto control valve assembly (44). Carefully separate valve spool parts, keeping shims (35) in their respective locations.

Check spool (39) and shift cover bore for damaged lands. Check tension of springs (36, 40 and 42) as follows:

Spring (36)—
  Free length . . . . 48 mm
  (1.91 in.)
  Test length . . . . 34 mm
  (1.33 in.)
  Test load . . . . 90-110 N
  (20-25 lbs.)
Spring (40)—
  Free length . . . . 22 mm
  (0.875 in.)

Test length . . . . 16 mm
(0.625 in.)
Test load . . . . 55-65 N
(12-15 lbs.)

Spring (42)—
  Free length . . . . 21 mm
  (0.810 in.)
  Test length . . . . 14 mm
  (0.540 in.)

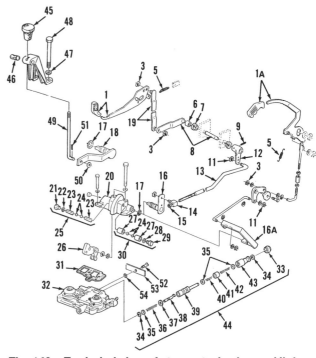

Fig. 149—Exploded view of pto control valve and linkage. Items (1 through 16) are used on tractors with Sound Gard Body. Items (18 and 45 through 54) are used on tractors without Sound Gard Body.

1. Control lever	
1A. Control lever	29. Plug
3. Snap ring	30. Hydraulic detent
5. Spring	31. Gasket
6. Bushing	32. Shift cover
7. Bushing	33. Plug
8. Lever shaft	34. Snap ring
9. Spring pin	35. Shims
11. Snap ring	36. Spring
12. Lever	37. Washer
13. Rod	38. Special pin
14. Swivel	39. Spool
15. Nut	40. Spring
16. Control lever shaft	41. Sleeve
16A. Control lever shaft	42. Spring
17. "O" ring	43. Actuator
18. Control lever shaft	44. Control valve spool assy.
19. Link	45. Knob
20. Cover	46. Pin
21. Plug	47. Bracket
22. "O" ring	48. Cap screw
23. Springs	49. Rod
24. Detent balls	50. Washer
25. Mechanical detent	51. Cotter pin
26. Control arm	52. Countersunk screw
27. Pistons	53. Retaining plate
28. "O" ring	54. Washer

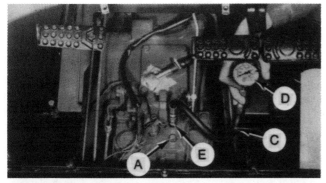

Fig. 148—Use a 2070 kPa (300 psi) test gage to check pto clutch and brake pressure.

A. Pto clutch test port
C. Test hose
D. Test gage
E. Brake test port

Test load . . . . 200-240 N
(45-55 lbs.)

When reassembling, determine correct shim packs (35) as follows: Install washer (37), spring (36) and one shim (35) and snap ring (34) on special pin (38). Place special pin (snap ring end first) on a spring scale. Press down on washer (37) with your fingers. The washer should break free of head on special pin at 30-34 N (6.5-7.0 lbs.). If necessary, add shims until correct spring preload has been obtained.

NOTE: There must always be at least one shim installed.

Install four shims (35—Fig. 150), spring (40), sleeve (41), spring (42), three shims (35), actuator (43) and snap ring (34) on spool (39). Place large end of spool on spring scale and press with fingers on actuator. At a load of 57-61 N (12-13 lbs.) actuator should break free from snap ring. If not, add or remove shims until correct spring preload has been obtained. When actuator just breaks free from snap ring, clearance

between sleeve (41) and shoulder of spool (39) should be 1.5-2.0 mm (0.06-0.08 inch).

Use all new "O" rings and reassemble by reversing the disassembly procedure.

147. PTO CLUTCH, BRAKE AND INPUT GEARS. Separate tractor between clutch housing and transmission case as outlined in paragraph 114 or 123. Remove oil lines (7 and 8—Fig. 151), pull connector (11) out to the front and remove suction line (13). If so equipped, remove Hi-Lo shift unit, Creeper transmission or front-wheel drive clutch. Remove drive gear to clutch drum Allen screws. Use two of the screws in jack screw holes and remove drive gear. Remove snap ring (1) and pull clutch drum (2) from pto clutch shaft (10). Remove connector (6) and oil inlet lines (4 and 5). Remove cap screws (5—Fig. 152) and lift out oil manifold (6) with pto brake and pto clutch shaft (32). Remove seal rings (15) and place

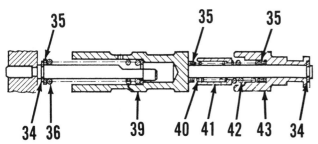

Fig. 150—Refer to text for adjustment of actuator and spool. Refer to Fig. 149 for legend.

Fig. 151—Installed pto clutch and oil lines typical of all models.

1. Snap ring
2. Clutch drum
3. Drive gear
4. Pto brake line
5. Pto clutch line
6. Pto brake inlet connector
7. Oil cooler line
8. Transmission oil filter line
9. Oil pump pressure line
10. Pto clutch shaft
11. Oil pump suction connector
12. Oil pressure line (front-wheel drive)
13. Oil pump suction line

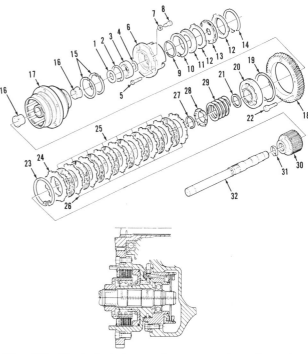

Fig. 152—Exploded view and cross section of pto clutch and brake assembly.

1. Spacer
2. Snap ring
3. Ball bearing
4. Snap ring
5. Cap screw (3)
6. Oil manifold
7. Washer
8. Allen screw
9. Seal ring
10. Seal ring
11. Piston
12. Brake plate
13. Brake disc
14. Snap ring
15. Seal rings
16. Bushings
17. Clutch drum
18. Drive gear
19. Seal ring
20. Piston
21. Seal ring
22. Allen screw
23. Snap ring
24. Back plate
25. Clutch discs
26. Clutch plates
27. Snap ring
28. Spring retainer
29. Coil spring
30. Clutch hub
31. Snap ring
32. Pto clutch shaft

clutch drum (17) in a protected vise with clutch side up. Remove snap ring (23) and lift out clutch hub (30), back plate (24), clutch plates (26) and clutch discs (25). Using a JDT-24A spring compressor or equivalent, compress spring (29) and remove snap ring (27). Remove retainer (28) and spring (29). Remove clutch piston (20) and seals (19 and 21). Remove snap ring (2) and withdraw pto clutch shaft (32) with ball bearing (3) from oil manifold (6). Remove snap ring (14), back plate (12), brake disc (13), brake plate (12), then remove brake piston (11) and seal rings (9 and 10).

Clean and inspect all parts and renew any showing excessive wear or other damage. Check spring (29) against the following specifications: Free length should be 98 mm (3.88 inches). Spring tension at a length of 46 mm (1.80 inches) should be 530-650 N (117-143 lbs.).

Use all new seal rings and reassemble by reversing the disassembly procedure, keeping the following points in mind. Tighten oil manifold to clutch housing cap screws (5) to 50 N·m (35 ft.-lbs.) torque. Install spring retainer (28) so that vanes are away from piston (20). Use Fig. 152 as a guide and install correct number of clutch discs (25) and plates (26). Eight plates and disks are used with dual speed pto, ten of each are installed on 540 rpm pto.

Reinstall assembly by reversing removal procedure.

148. OUTPUT SHAFT AND GEARS. To remove the output shaft and gears, first drain transmission case, then unbolt and remove bearing quill assembly (16 through 26—Fig. 153). Remove spring washer (15) and gear (14), then withdraw pto shaft and gear (10). Remove snap ring (28). Pull shaft (30) out of transmission case and remove thrust washers (31) and reduction gear (34) with bearing needles (32) and spacer (33).

Complete disassembly of removed components. To remove sleeve coupler (2) and needle bearing (5), tractor must be split between clutch housing and transmission case.

Clean and inspect all parts and renew any showing excessive wear or other damage. Use all new gasket,

oil seal and "O" rings and, using Fig. 153 as a guide, reinstall by reversing the removal procedure.

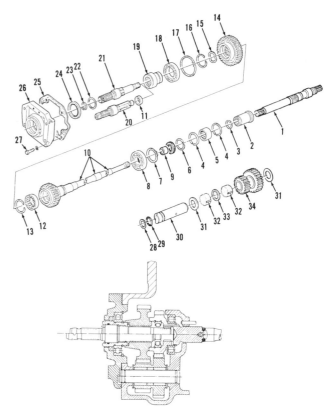

Fig. 153—Exploded view and cross section of dual speed pto rear shafts and gears.

1. Pto clutch shaft	18. Ball bearing
2. Sleeve coupler	19. Pilot
3. Snap ring	20. 540 rpm stub shaft
4. Snap rings	21. 1000 rpm stub shaft
5. Needle bearing	22. "O" ring
6. Snap ring	23. Snap ring
7. Snap ring	24. Oil seal
8. Ball bearing	25. Gasket
9. Bushing	26. Bearing quill
10. Pto shaft & gear	27. Cap screw
11. Bearing race	28. Snap ring
12. Roller bearing	29. "O" ring
13. Snap ring	30. Countershaft
14. Gear	31. Thrust washer
15. Spring washer	32. Bearing needles
16. Snap ring	33. Spacer
17. Snap ring	34. Reduction gear

FRONT POWER TAKE-OFF

Tractors may be equipped with a pto located at the front of the tractor that is driven by a shaft running through the hydraulic pump. The front pto operates at 1000 rpm and is engaged and disengaged by an electric solenoid/hydraulic valve that operates a multiple disc clutch. A mechanical jaw clutch disengages the drive shaft from gears when not in use.

All Models So Equipped

150. REMOVE AND REINSTALL. The front pto unit can be removed from the front of tractor as follows. Drain oil from unit by removing plug (1—Fig. 154) from the lower, center front of the housing. Remove shaft cover if so equipped and disconnect oil lines (Fig. 155) from pto housing. Disconnect the cable from lever (5—Fig. 154) and pull cable and housing free from pto unit. Secure lifting eyes and hoist to the unit and remove the four attaching screws (4). Pull front pto assembly forward away from tractor.

When installing, carefully guide the drive shaft into the connecting sleeve at hydraulic pump. Tighten the four retaining screws (4) to 400 N·m (300 ft.-lbs.) and the drain plug to 30 N·m (23 ft.-lbs.) torque. Fill unit with transmission fluid and check for leaks, especially at tubing connections.

151. OVERHAUL. Most service to the front pto unit will necessitate removal of the unit as outlined in paragraph 150. Remove bearing quills (6 and 11—Fig. 154), being careful not to lose shims (8 and 15). Unbolt and remove the front half of the pto housing

(2), remove plug and detent assembly (17), then lift out the input shaft (23) with drive gear (25), engagement collar (24) and bearings (9, 26 and 27). The intermediate gear (39), shaft (41) and bearings (38) can be removed from rear half of housing after removing retaining screw (36) and washer (37). Unbolt and remove oil baffle (44), then lift pto clutch and output shaft assembly (43) from rear housing and oil manifold. Oil manifold and brake assembly and control valve can be unbolted from housing.

Remove output shaft (81—Fig. 156), hub (84) and associated parts. Remove backing plate (86), internally splined clutch friction discs (78) and plates (79) with external tabs. Use a special tool as shown in Fig. 157 to compress spring (75—Fig. 156), then remove snap ring (77). Compressed air can be used to blow piston (73) from bore. Be careful to prevent ring (72)

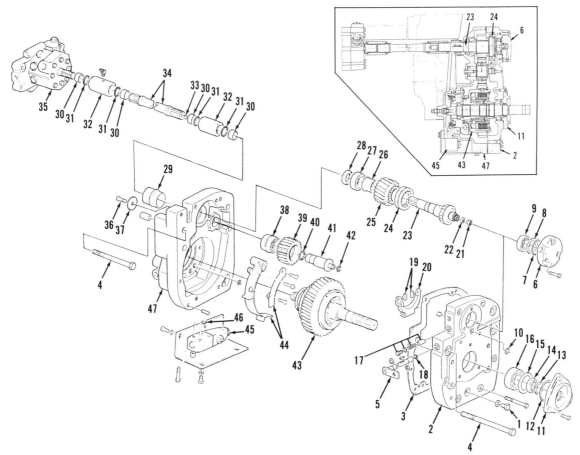

Fig. 154—Partially exploded view of front pto. The four larger screws (4) attach the unit to tractor.

1. Drain plug	13. Seal ring	25. Input gear	37. Washer
2. Housing front half	14. Seal	26. Bushing	38. Bearing assy.
3. Gasket	15. Shims	27. Tapered roller bearing	39. Intermediate gear
4. Attaching screws	16. Tapered roller bearing	28. Seal	40. "O" ring
5. Shift lever	17. Detent assy.	29. Plastic bushing	41. Intermediate shaft
6. Bearing quill	18. "O" ring	30. Spacer	42. "O" ring
7. "O" ring	19. Guide shoes	31. "O" ring	43. Clutch & output shaft
8. Shims	20. Shift fork	32. Coupling	44. Baffle
9. Tapered roller bearing	21. Seal ring holder	33. Spring	45. Control valve
10. Seal ring	22. Seal ring	34. Drive shaft	46. Seal rings (4 used)
11. Bearing quill	23. Input shaft	35. Hydraulic pump	47. Housing rear half
12. "O" ring	24. Coupling	36. Cap screw	

from catching in groove in drum (68) for snap ring (85). Remove snap ring (49), brake plate (50), disc (51) and plate (52), then use air to blow piston (53) from bore in manifold (63).

Inspect parts and renew any that are grooved, cracked, heat discolored or otherwise questionable.

Clutch plates (79),
thickness new. . . . . 2.3 mm
(0.09 inch)
Clutch disc (78),
thickness new. . . . . 2.44-2.54 mm
(0.096-0.100 inch)
Brake disc (51),
thickness new. . . . . 5.9-6.1 mm
(0.232-0.240 inch)
Clutch spring (75)—
Free length. . . . . 98 mm
(3.88 inches)
Tension at 46 mm . . . . 530-650 N
(1.80 inches)   (117-143 lbs.)
Modulation spring (60)—
Free length. . . . . 87 mm
(3.42 inches)
Tension at 48 mm . . . . 53-65 N
(1.88 inches)   (12-14.5 lbs.)
Modulation spring (61)—
Free length. . . . . 54 mm
(2.13 inches)

Tension at 31 mm. . . . . 123-151 N
(1.22 inches)   (27.5-40 lbs.)

Assemble in reverse of disassembly procedure. Make sure that the three seal rings (48) are in place in rear housing (47) before installing the manifold and brake assembly, then tighten retaining screws to 55 N•m (40 ft.-lbs.) torque. Eight clutch plates (79) and eight discs (78) should be alternated, beginning with a plate (79) with external tangs and ending with a disc (78) with internal splines. Backing plate (86) should be installed with flat side toward last friction disc (78). Screws attaching oil baffle (44—Fig. 154) should be coated with Loctite 242 to prevent loosening. Tighten screw (36) retaining shaft (41) to 75 N•m (55 ft.-lbs.) torque. Tighten screws holding halves of housing (2 and 47) together to 55 N•m (40 ft.-lbs.) and drain plug to 30 N•m (23 ft.-lbs.) torque. Bearings (9 and 27) and bearings (16—Fig. 154 and 64—Fig. 156) should be adjusted to have 0.0-0.05 mm (0.0-0.002 inch) preload, by the addition of shims (8 and 15—Fig. 154). To determine the thickness of shims to install, push bearing races in to remove all play from bearings, then measure depth of bearing cup as shown at (B—Fig. 158). Measure the height of lip on appropriate quill as shown at (C). Subtract the height of the lip (C) from the depth of bearing (B), then add the desired preload to this difference. Install shims (8 or 15—Fig. 154) equal to the calculated thickness. If removed, fitting (H—Fig. 155) should be installed

Fig. 155—View of oil lines to front pto.

A. Leakage oil line
B. Lubrication oil line
C. Pressure oil line
D. Lubrication oil line
E. Pto clutch test port plug
F. Return line
G. Pto brake test port plug
H. Elbow fitting for return line

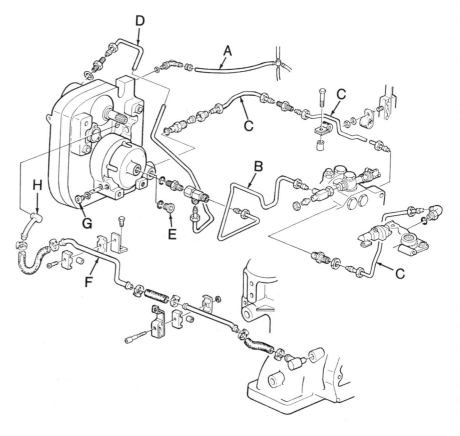

with a coating of Loctite 242 on the lip pressed into housing.

Refer to Fig. 159 for exploded view of solenoid valve used on front-mounted pto.

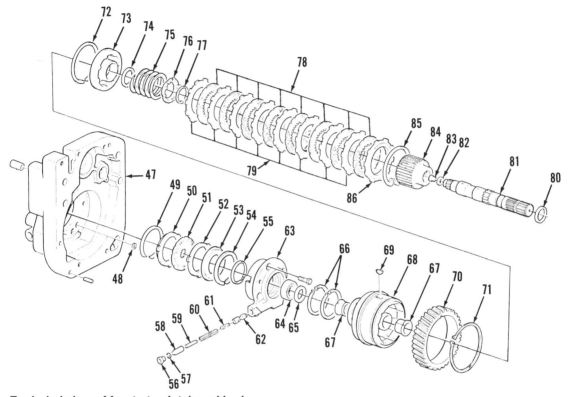

Fig. 156—Exploded view of front pto clutch and brake.

47. Housing rear half	57. "O" ring	67. Bushings	77. Snap ring
48. Seal rings (3 used)	58. Piston	68. Clutch drum	78. Clutch friction discs
49. Snap ring	59. Pin	69. Woodruff key	79. Clutch plates (external lugs)
50. Brake plate	60. Modulator spring (long)	70. Ring gear	80. Back-up washer
51. Brake disc (internal splines)	61. Modulator spring (short)	71. Snap ring	81. Output shaft
52. Brake plate	62. Modulation valve	72. Seal ring	82. Seal ring
53. Brake piston	63. Oil manifold	73. Clutch piston	83. Seal ring holder
54. Seal ring	64. Tapered roller bearing	74. Seal ring	84. Clutch hub
55. Seal ring	65. Spacer washer	75. Clutch release spring	85. Snap ring
56. Plug	66. Seal rings	76. Spring plate	86. Backing plate

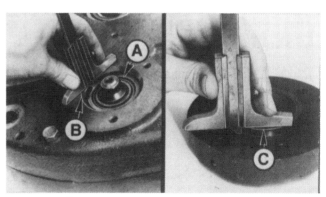

Fig. 157—Use special tool JDT-24A or equivalent to compress spring so that snap ring can be removed or installed.

Fig. 158—Refer to text and measure as shown to determine correct thickness of shims to be installed at (8 and 15-Fig. 154). Bearing cup is shown at (A).

Fig. 159—Exploded view of solenoid valve used to operate front pto. Smaller cam of spool (7) should be toward end of housing (6) with spring pin (10).

1. Flange
2. Cap
3. "O" rings
4. Springs
5. Centering discs
6. Housing
7. Spool
8. Iron core
9. "O" rings (4 used)
10. Spring pin
11. "O" ring
12. Solenoid
13. "O" ring
14. Slotted nut

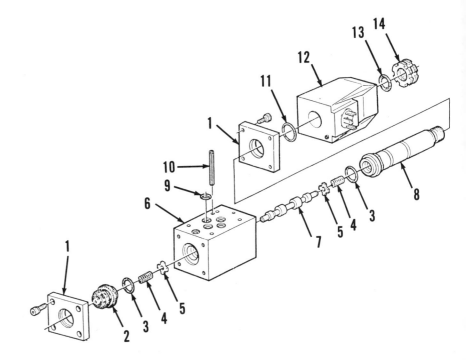

HYDRAULIC LIFT SYSTEM

The hydraulic lift system is a closed-center, constant-pressure-type. The standby pressure of 15,800-16,200 kPa (2300-2350 psi) is furnished by an eight piston, constant running variable displacement pump. Two sizes of main hydraulic pumps are available: a small pump displaces 22.6 cm3 (1.38 cubic inches displacement) and a larger pump displaces 40 cm3 (2.4 cubic inches displacement). Pump assemblies and internal parts are different; however, operation of the two pumps is basically the same.

Pump is mounted in tractor front support and is driven by a coupling from front end of engine crankshaft. Tractors equipped with a front pto, have a hydraulic pump with a through shaft that also drives the front pto unit. Charging oil for the hydraulic main pump is supplied by the transmission oil pump and oil not used by the main pump is routed to the auxiliary hydraulic oil reservoir. This reservoir provides an auxiliary supply of oil when transmission oil pump is unable to meet the demand of the main pump. When there is little or no demand by the main pump, the overflow from auxiliary reservoir is returned to the clutch housing through an oil cooler where part of it fills the brake control valve reservoir. The remainder lubricates the transmission shafts and gears.

TROUBLE-SHOOTING

All Models

152. Following are symptoms, and their possible causes, that may occur during operation of the hydraulic lift system. By using this information in conjunction with the Test and Adjust information, no trouble should be encountered in servicing the hydraulic system.

1. Slow system operation. Could be caused by:
   a. Clogged transmission oil filter.
   b. Transmission oil pump inlet screen plugged.
   c. Faulty transmission oil pump.
   d. Transmission oil pump relief stuck open.
   e. Main hydraulic pump stroke control valve not seating properly.
   f. Oil leak on low pressure side of system.
   g. Hydraulic pump shut-off screw turned in.
   h. Crankcase outlet valve seized.

2. Erratic pump operation. Could be caused by:
   a. Pump stroke control valve not seating properly.
   b. Leaking pump inlet or outlet valves or valve "O" rings.
   c. Broken or weak pump piston springs.

3. Noisy pump. Could be caused by:
   a. Worn drive parts or loose cap screws in drive coupling.
   b. Air trapped in oil cavity of pump stroke control valve.

4. No hydraulic pressure. Could be caused by:
   a. Pump shut-off valve closed.
   b. No oil in system.
   c. Faulty pump.

5. Rockshaft fails to raise or raises slowly. Could be caused by:
   a. Excessive load.
   b. Low pump pressure or flow.
   c. Rockshaft piston "O" ring faulty.
   d. Flow control valve maladjusted.
   e. Surge relief valve defective.
   f. Cam follower adjusting screw maladjusted.
   g. Transmission oil filter plugged.
   h. Defective seals between cylinder and rockshaft housing or between rockshaft housing and transmission case.

6. Rockshaft settles under load. Could be caused by:
   a. Leaking discharge valve.
   b. Leaking rockshaft cylinder check valve.
   c. Leaking cylinder end plug.
   d. Faulty rockshaft cylinder valve housing.
   e. Surge relief valve leaking.

7. Erratic action of rockshaft control valves. Could be caused by:
   a. Control valves maladjusted.
   b. Rockshaft piston "O" ring faulty.
   c. Discharge valve leaking.
   d. Surge relief valve leaking.

8. Rockshaft lowers too fast or too slow. Could be caused by:
   a. Rate-of-drop screw maladjusted.
   b. Valve linkage damaged.

9. Rockshaft raises too fast. Could be caused by:
   a. Flow control valve incorrectly set.

10. Insufficient load response. Could be caused by:
    a. Control valve clearance excessive.
    b. Control valves sticking.
    c. Control lever not positioned correctly on quadrant.
    d. Worn load control shaft or bushings.
    e. Negative stop screw turned in too far.

11. Hydraulic oil overheating. Could be caused by:
    a. Control valves adjusted too tight and held open.
    b. Control valves leaking.
    c. Control valve "O" rings faulty.
    d. Surge relief valve leaking.
    e. Oil cooler plugged.

HYDRAULIC SYSTEM TESTS

Before making any tests on the main hydraulic pump or lift system, be sure the transmission oil pump is satisfactory because the performance of the main pump is dependent upon being charged by the transmission oil pump. For information on testing the transmission oil pump, refer to paragraph 120 for Collar Shift transmission models or paragraph 127 for Synchronized transmission models.

All Models

153. MAIN HYDRAULIC PUMP TEST. The main hydraulic pump should be tested for standby pressure as described in paragraph 155 before testing the rate of flow. Low standby pressure will result in low flow. Refer to paragraph 154 for flow testing procedure using hydraulic test unit after making sure that standby pressure is correct.

154. When testing the rate of flow of the main hydraulic pump using the hydraulic test unit, attach the unit as follows. Disconnect outlet line from pump in front of pressure control valve and connect the flow meter inlet line as shown in Fig. 160. Attach outlet from tester to the filter cover as shown or to the right side port of the left hand quick coupler. If selective control valve port is used, the handle of the left hand selective control valve must be secured in the rearward position during test.

Start engine and operate until transmission oil is heated to 65° C (150° F), then set engine speed at 2000 rpm. Close tester control valve so tester pressure gage reads 13,800 kPa (2000 psi) and observe rate of flow, which should be as follows:

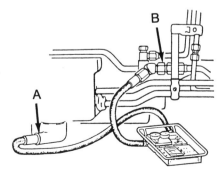

Fig. 160—A flow control and pressure test unit can be attached as shown in place of fitting that supplies system pressure to power steering system.

23 cm³ (1.38 cu. in.) pump . . . . 34 L/m
(9 gpm)

40 cm^3 (2.4 cu. in.) pump................. 68 L/m
(18 gpm)

If main hydraulic pump will not meet both pressure and flow requirements, it must be removed and overhauled as outlined in paragraph 163.

155. Test main hydraulic pump standby pressure by installing a 35,000 kPa (5000 psi) test gage as shown in Fig. 162.

Start engine and warm transmission oil to a temperature of 65° C (150° F). With engine running at 2000 rpm and transmission in NEUTRAL, move left selective control valve lever forward and note gage reading, which should be 15,900-16,200 kPa (2300-2350 psi). If pump standby pressure is incorrect, refer to Fig. 161, loosen jam nut and turn pump stroke control valve adjusting screw in to increase or out to decrease pressure. If pump will not produce the cor-

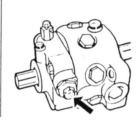

Fig. 161—Arrow indicates locations of pump stroke control valve adjuster screw and locknut.

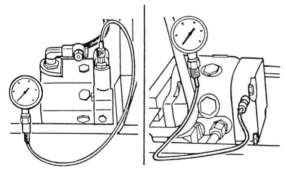

Fig. 162—View of gage attached for checking main pump standby pressure.

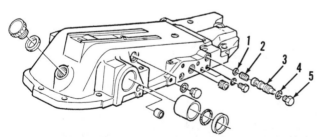

Fig. 163—View showing location of flow control valve in rockshaft housing.
1. Washers
2. Spring
3. Flow control valve
4. Seal ring
5. Plug

rect standby pressure or if pressure pulsates more than 690 kPa (100 psi), the following are possible causes:
  a. Gage improperly dampened.
  b. Stroke control valve stuck or held open.
  c. Crankcase valve open.
  d. Pumping pistons scored.
  e. Inlet or discharge valves leaking.

Refer to paragraph 163 for overhauling pump.

156. FLOW CONTROL VALVE TEST. A flow control valve is incorporated into the rockshaft valve circuit to reduce the oil flow of some models. See Fig. 163. Check the flow control valve as follows: Operate the engine at 2000-2300 rpm, move the selective control valve lever forward to raise the rockshaft and measure the time it takes to raise the rockshaft. It should take 2-3 seconds to raise from the lowest to the highest position. If flow does not raise the rockshaft within 2-3 seconds, refer to Fig. 163, remove plug, valve and spring, then vary washers (shims) as necessary. Also, check spring, which should test 55-65 N (11.7-14.3 lbs.) when compressed to a length of 20 mm (0.79 inch).

HYDRAULIC SYSTEM ADJUSTMENTS

The following paragraphs outline adjustments that can be made when necessary to correct faulty hydraulic operation, or that must be made when reassembling a hydraulic lift system that has been disassembled for service.

However, because of the interaction of component parts, the adjustments must be made in the following order:
  1. Negative stop screw.
  2. Rockshaft control lever neutral range.
  3. Control lever position.
  4. Load control.
  5. Rate-of-drop.

All Models

157. NEGATIVE STOP SCREW. The negative stop screw is shown in Fig. 164 and is used to provide a stop for the load control arm. To adjust, loosen jam nut and turn stop screw in until it just contacts the load control arm, then turn screw out the specified amount and tighten lock nut. Contact of stop screw with load control arm can be more easily felt if rockshaft housing filler plug is removed and a screwdriver is held against upper end of the load control arm. After screw just contacts arm, back screw out 1/8 turn on 2750 models, 1/3-1/2 turn on all other models. It is important that this is correctly adjusted because it can adversely affect remaining system adjustments.

158. CONTROL LEVER NEUTRAL RANGE.
To adjust the control lever neutral range, load lift links with a weight of approximately 100 kg (220 lbs.). Remove the pipe plug located directly in front of control lever tube, which will expose the control valve adjusting screw (1—Fig. 165 or Fig. 166). The adjusting screw is located under the Sound Gard Body of models so equipped. Move load selector lever (2) to "MIN" depth position and lower the lift arms fully. Move the rockshaft control lever (3) to center of quadrant or console and close the rate-of-drop screw. Start engine and run at 1300 rpm. When engine starts, turn adjusting screw (1) counterclockwise until the arms just start to raise, then turn adjusting screw (1) ¼ turn clockwise. Open the rate-of-drop screw. Move the control lever (3) forward until rockshaft just starts to lower and mark this point on edge of quadrant or console. Now move lever slowly rearward until rockshaft just begins to raise and mark this point on edge of quadrant or console. Distance between these two marks should be as follows:

2750—
    W/o Sound Gard Body . . . . 5 mm
                                     (0.2 inch)
    With Sound Gard Body . . . . 15 mm
                                       (0.6 inch)
2755, 2955—
    W/o Sound Gard Body . . . . 2-4 mm
                                  (0.08-0.16 inch)
    With Sound Gard Body . . . . 12-15 mm
                                  (0.5-0.6 inch)
2855N . . . . 3-6 mm
                                  (0.12-0.24 inch)

If distance between marks is not correct, turn the control valve adjusting screw (1) clockwise to increase or counterclockwise to decrease control lever neutral range.

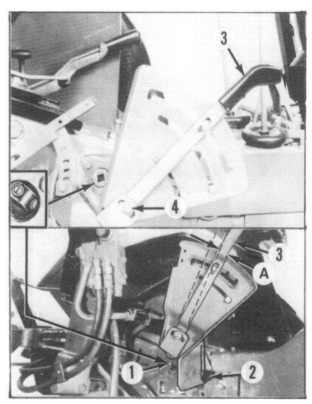

Fig. 165—Refer to text for adjustment of control lever neutral range.
1. Adjusting screw
2. Load selector lever
3. Rockshaft control lever
4. Adjusting nut

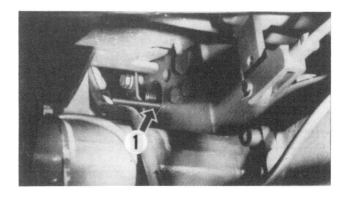

Fig. 164—View of negative stop screw and lock nut.

Fig. 166—View showing adjustment of control lever position on models without Sound Gard Body. Refer to text.

159. CONTROL LEVER POSITION. To adjust control lever position on models without Sound Gard Body, operate engine at 1300 rpm and place load selector (2—Fig. 166) in lower "MIN" position. Then, move rockshaft control lever (3) fully forward to completely lower rockshaft. Loosen control lever adjusting nut (4—Fig. 165) and move control lever rearward until there is the correct amount of clearance at (A) between control lever and end of quadrant slot. Rotate control shaft arm counterclockwise to the point where rockshaft just starts to raise. Tighten control lever nut (4). Correct clearance (A) is 10-25 mm (0.4-0.8 inch) for 2750 and 2855N models; 10-13 mm (0.39-0.51 inch) for 2755 and 2955 models. Pull lever (3) to the rear and check to be sure that a small gap exists at the rear.

On models equipped with Sound Gard Body, remove side cover from console. Pull back upholstery, remove plastic rivets and remove plastic cover from right lift arm. Place load selector lever in "MIN" position. With engine operating at 1300 rpm, move control lever forward until rockshaft is fully lowered; then, move control lever to the rear until its front edge is aligned with position "7-7.5" on console. See Fig. 167. Working through the uncovered opening, adjust position of rod end (B—Fig. 168) on link (A) until rockshaft starts to raise. Tighten locknuts on turnbuckle. Move control lever rearward until its front edge is aligned with "1.5-2.5" position. Rockshaft

should now raise to its highest position. If incorrect, readjust at "7-7.5" position.

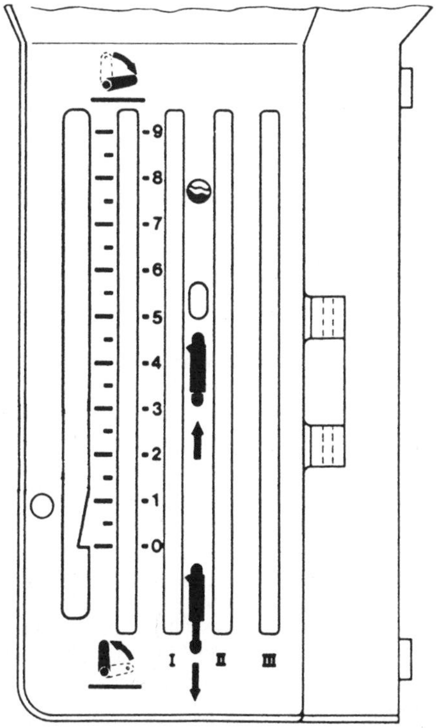

160. LOAD CONTROL ARM. On models without Sound Gard Body, remove rockshaft filler hole plug to expose the load control arm cam follower adjusting screw (5—Fig. 169). Place load selector lever (2—Fig. 166) in "MAX" position. With engine operating at 1300 rpm, move control lever (3) fully forward to completely lower rockshaft. Then, move control lever rearward until distance between control lever and end of quadrant slot is correct. Distance should be 85-95 mm (3.34-3.74 inches) for 2750 models; 47-53 mm (1.88-2.12 inches) for 2755 and 2955 models; 85-100 mm (3.34-3.94 inches) for 2855N model. Hold control lever in this position, loosen jam nut and turn cam follower adjusting screw (5—Fig. 169) counterclockwise until rockshaft is fully lowered. Then, turn adjusting screw clockwise until rockshaft starts to raise. Tighten jam nut. A special tool (JDG402) is available to facilitate adjustment.

NOTE: If rockshaft begins to raise before control lever reaches the recommended setting, it will be necessary to turn cam follower adjusting screw counterclockwise to allow the lever to be positioned while rockshaft remains in lowered position.

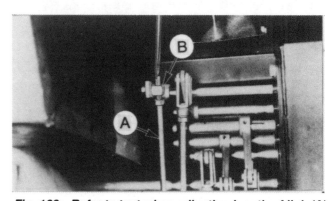

Fig. 168—Refer to text when adjusting length of link (A) by repositioning end (B).

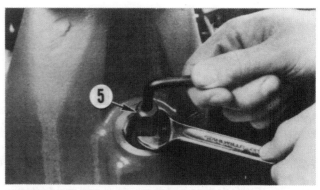

Fig. 169—View showing location of load control arm cam follower adjusting screw (5). Remove rockshaft filler plug for access.

Fig. 167—Top view of hydraulic control console on models with Sound Gard Body.

Illustrations for Fig. 167, Fig. 168 and Fig. 169 reproduced by permission of Deere & Company. Copyright Deere & Company.

On models equipped with Sound Gard Body, remove rockshaft filler plug. Place load selector lever in "MAX" position. With engine operating at 1300 rpm, move control lever forward until rockshaft is fully lowered. Move control lever rearward until its front edge is aligned with position "3.5" for 2750 models, "2-2.5" for 2755 and 2855N models or "3.5" for 2955 models. Refer to Fig. 167. If rockshaft starts to raise before control lever reaches the correct position, loosen jam nut and turn cam follower adjusting screw (5—Fig. 169) counterclockwise to allow lever to be positioned while rockshaft remains in lowered position. Turn adjusting screw (5) clockwise until rockshaft just starts to raise, then tighten jam nut.

161. RATE-OF-DROP. The rate-of-drop adjusting screw is located on top side of rockshaft housing. Fig. 170 shows adjusting screw with handle (6) for models with and without Sound Gard Body.

On all models, turn adjusting screw clockwise to decrease rate-of-drop or counterclockwise to increase rate-of-drop. Tighten jam nut after adjustment is completed. Rate-of-drop will vary with the weight of the attached implement. Rate-of-drop time should not be less than two seconds. However, adjusting

Fig. 170—View of rate-of-drop adjustment screw (6) for models with and without Sound Gard Body. Refer to text.

screw should not be turned out more than one complete turn.

MAIN HYDRAULIC PUMP

All Models

162. A 23 cc (1.38 cu. in.) or a 40 cc (2.40 cu. in.) piston-type pump is used. Pumps with a drive shaft that continues through pump are available to drive the front pto unit.

Pump should maintain about 14,000 kPa (2050 psi) working pressure with a standby pressure of 15,900-16,200 kPa (2300-2350 psi). Standby operation of the pump occurs when pressure in the pump crankcase builds high enough to hold the pump pistons away from the pump cam. Pump crankcase (standby) pressure is controlled by the stroke control valve located in a bore in the pump valve housing.

A pump shut-off (destroking) screw may be optionally available that will make pump inoperative and will act as an aid during cold weather starts.

163. R&R AND OVERHAUL. Remove grille screens, disconnect light bar wires at connector and remove hood. Drain fuel from main fuel tank, lines and remote tank, then disconnect lines from fuel tank. Remove air cleaner and air conditioning condenser if so equipped. Remove fuel tank brackets and lift fuel tank from tractor. On models with through drive, it is necessary to remove the radiator. On all models, disconnect oil lines at pump and immediately plug or cap all openings to prevent loss of oil and dirt from entering system. Remove the four screws (4—Fig. 171 or 4A—Fig. 172) and loosen the clamp screw (2—Fig. 171) or set screws (2A—Fig. 172). Slide pump drive adapter (3, 3H, 3P or 3PH—Fig. 171 or Fig. 172) forward on pump shaft. Remove drive cushion (6, 6P

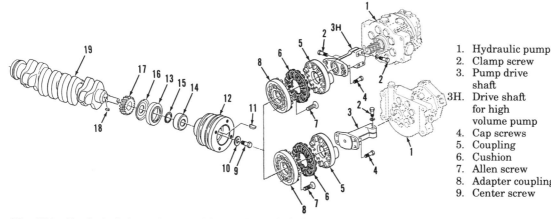

1. Hydraulic pump	
2. Clamp screw	
3. Pump drive shaft	10. Washer
3H. Drive shaft for high volume pump	11. Shaft key
	12. Pulley
	13. Oil seal
4. Cap screws	14. Seal sleeve
5. Coupling	15. "O" ring
6. Cushion	16. Oil slinger
7. Allen screw	17. Crankshaft timing gear
8. Adapter coupling	18. Drive key
9. Center screw	19. Late crankshaft
	19E. Early crankshaft

Fig. 171—Exploded view of pump drive and crankshaft used on models without front pto. Note differences for models with standard- and high-volume hydraulic pumps.

or 6PH) and front coupler half (5—Fig. 171). Unbolt and remove pump assembly.

164. OVERHAUL 23 cc (1.38 cu. in.) PUMP. Thoroughly clean exterior of pump and check pump shaft end play before disassembling. Use a dial indicator to accurately measure pump shaft end play and record measurement for aid in reassembling. End play should be 0.1-0.9 mm (0.004-0.035 inch) for models without through shaft, 0.025-0.10 mm (0.001-0.004 inch) for models with through drive shaft for the front pto. If end play is excessive, install new thrust washers (21—Fig. 173) when assembling models without through shaft. To reduce end play on models with through shaft, add shims (57) as necessary when assembling to reduce end play to within limits. Bearing wear or wrong number of adjusting shims can cause excessive end play.

Clamp pump in a vise and remove cover (2), then remove crankcase outlet valve (34, 40, 43 and 49) from pump housing. Remove outer thrust washer (21).

> NOTE: At this time, shaft (23) can be removed if desired by removing Woodruff key and pushing shaft out of body and cam ring (22). Be sure not to lose any of the 33 rollers (24) that will be loose. However, if pump requires disassembly, it is a good policy to completely disassemble pump and inspect all parts and to remove pump pistons before removing pump shaft.

Remove piston assemblies (20, 46, 47 and 48), then, if not already removed, remove shaft (23), rollers (24) and cam ring (22). Remove inlet valves (8, 10, 11 and 12) and outlet valves (14, 15, 53, 17, 18 and 20). Identify all pistons, valves, springs and seats so they can be reinstalled in their original positions. Remove plug (42) and filter screen (45). Loosen jam nut and remove stroke control valve adjusting screw (29), spring (31), spring guide (52) and valve (32).

Clean and inspect all parts. Use the following specifications as a guide for renewal of parts:

Thrust washer (21) thickness . . . . 2.21-2.31 mm
(0.087-0.091 inch)
Piston bore ID . . . . 17.27-17.30 mm
(0.680-0.681 inch)
Piston (46) OD . . . . 17.25-17.27 mm
(0.679-0.680 inch)
Cam race ID. . . . . 45.72-45.75 mm
(1.800-1.801 inch)
Pump shaft (23) cam OD . . . . 37.77-37.79 mm
(1.487-1.488 inch)

Discard all old "O" rings and use new during assembly. Pay particular attention to shaft bore around the quad ring seal groove as leakage at this point can cause the pump to be slow in going out of stroke. If outlet valve seats (14) are damaged, drive them out and install new seats with large chamfered end toward bottom of bore. Press seats in to 29.75 mm (1.171 inch) below top face surface of bore using special tool JDH-39-1 or equivalent. OD of new outlet valve (15) is 1.54-1.55 mm (0.609-0.611 inch) and any valve that is distorted, scored or worn should be renewed. If stroke control valve seat (33) is damaged, remove plug (20) or shut-off assembly (35) if so equipped and drive out seat (33). Install new valve chamfered end first and drive it in bore until it bottoms. Spring (31) should test 710-850 N (160-190 lbs.) when compressed to a length of 63.5 mm (2½ inches). Renew any pistons (46) that are scored or pitted. Piston springs (48) should exert 80-100 N (18-22 lbs.) when compressed to 32 mm (1.26 inches) and springs must test within 7 N (1.5 lbs.) of each other when compressed to test length of 32 mm (1.26 inches).

When reassembling, dip all parts in oil. Use grease to hold rollers in ID of cam ring during assembly. Seal (26) is installed with printed side toward outside.

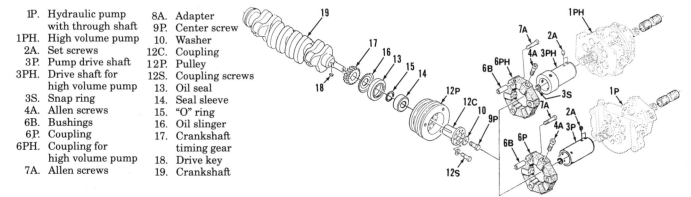

1P. Hydraulic pump with through shaft
1PH. High volume pump
2A. Set screws
3P. Pump drive shaft
3PH. Drive shaft for high volume pump
3S. Snap ring
4A. Allen screws
6B. Bushings
6P. Coupling
6PH. Coupling for high volume pump
7A. Allen screws
8A. Adapter
9P. Center screw
10. Washer
12C. Coupling
12P. Pulley
12S. Coupling screws
13. Oil seal
14. Seal sleeve
15. "O" ring
16. Oil slinger
17. Crankshaft timing gear
18. Drive key
19. Crankshaft

Fig. 172—Exploded view of pump drive and crankshaft used on models with front pto. Note differences for models with standard- and high-volume hydraulic pumps.

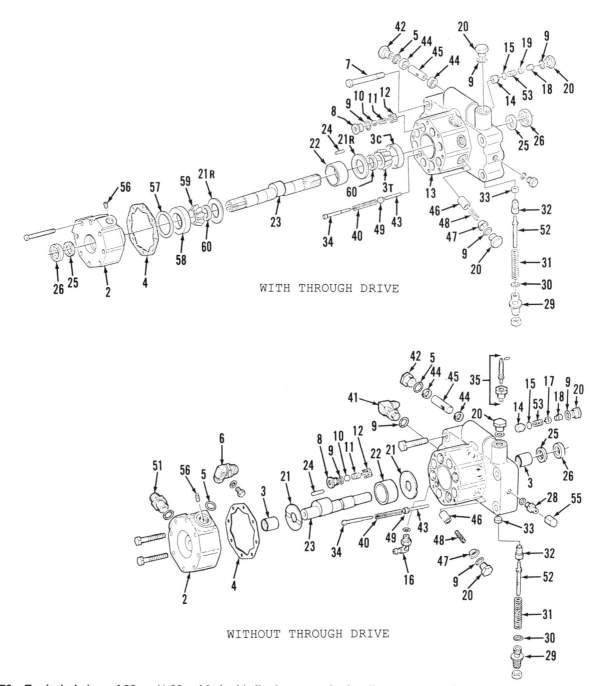

WITH THROUGH DRIVE

WITHOUT THROUGH DRIVE

Fig. 173—Exploded view of 23 cc (1.38 cubic inch) displacement hydraulic pumps used on some tractors.

2. Pump cover	14. Outlet valve seat	28. Connector
3. Needle bearings	15. Outlet valve	29. Adjusting screw
3C. Bearing cup	16. Elbow	30. "O" ring
3T. Tapered roller bearing	17. Guide	31. Spring
4. Gasket	18. Stop	32. Stroke control valve
5. "O" rings	19. "O" ring	33. Valve seat
6. Elbow	20. Plug	34. Spring guide
7. Screws	21. Thrust washer	35. Pump shut-off assy.
8. Inlet valve seat	21R. Thrust washer	40. Spring
9. "O" ring	22. Cam race	41. Elbow
10. Inlet valve ball	23. Pump shaft (cam)	42. Plug
11. Spring	24. Bearing rollers (33)	43. Pin
12. Guide	25. Quad ring	44. Seals
13. Housing	26. Oil seal	45. Filter screen

46. Pump piston	
47. Protective sheath	
48. Piston spring	
49. Crankcase outlet valve	
51. Connector	
52. Valve guide	
53. Spring	
55. Cap	
56. Orifice	
57. Spacer	
58. Bearing cup	
59. Tapered roller bearing	
60. Spacer	

112

Thrust washers are installed with grooved side away from pump shaft cam.

Cover to body screws—
First torque . . . . 20 N·m
(14 ft.-lbs.)
Final torque . . . . 50 N·m
(35 ft.-lbs.)
Torque, piston plugs (20). . . . . 125 N·m
(90 ft.-lbs.)
Torque, pump mounting screws . . . . 120 N·m
(85 ft.-lbs.)

Reinstall pump by reversing the removal procedure and adjust standby pressure as outlined in paragraph 155.

165. OVERHAUL 40 cc (2.4 cu. in.) PUMP. Thoroughly clean exterior of pump and check pump shaft end play before disassembling. Use a dial indicator to accurately measure pump shaft end play and record measurement for aid in reassembling. End play should be 0.025-0.10 mm (0.001-0.004 inch) for all models. End play is adjusted by adding or removing shims (26—Fig. 175). Bearing wear or wrong number of adjusting shims can cause excessive end play.

To disassemble the pump, remove the four cap screws retaining stroke control housing (15—Fig. 176) to front of pump and remove housing. Mark pump housing and place removed parts in a compartmented tray so parts will be reinstalled in their original locations. Withdraw outlet stops, guides, springs and valves (28 through 31—Fig. 175). Be especially careful to mark the pistons and bores so that each can be reinstalled in same bore from which it was removed. Remove all piston plugs (35), springs (34) and pistons (33), then carefully withdraw pump shaft (22) with bearing cones (19 and 24), thrust washers (21), spacers (20), cam race (23) and bearing needles (27).

Remove plug (7) retaining inlet valve assembly (5) and check inlet valve lift using a dial indicator. Lift should be 2.0-3.0 mm (0.080-0.120 inch). If lift exceeds 3 mm (0.120 inch), spring retainer is probably worn and valve should be renewed. Also check for apparent excessive looseness of valve stem in guide. Do not remove inlet valve assembly unless renewal is indicated or outlet valve seat (32) must be renewed. Do not reinstall a removed inlet valve (5) because they depend on their tight press fit for sealing. To remove an inlet valve, use a small pin punch and drive valve out, working through outlet valve seat (32). Outlet valve seat can be driven out after inlet valve is removed. Outlet valve spring (30) should test 11.3-14.0 N (2.5-3.1 lbs.) when compressed to a length of 7.6 mm (0.30 inch).

All eight piston springs (34) should be of the same color and should test within 7 N (1.5 lbs.) of each other when compressed to length of 41 mm (1.62 inches). Piston springs should test as follows when compressed to 41 mm (1.62 inches):

Yellow. . . . . 151-158 N
(34.0-35.5 lbs.)
Green . . . . 158-165 N
(35.5-37.0 lbs.)
Blue . . . . 165-171 N
(37.0-38.5 lbs.)
Red . . . . 171-178 N
(38.5-40.0 lbs.)

Inspect bearings (19 and 24), spacers (20), thrust washers (21), shaft (22), race (23) and bearing needles (27) for excessive wear or other damage and renew as necessary.

When reassembling pump, install new seal ring (10) in housing on early-style pump and on shaft on late pump. Install oil seal (9) only deep enough to allow snap ring to enter groove, to avoid blocking the relief hole in body.

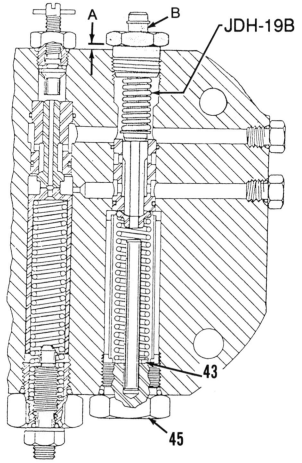

Fig. 174—Special adjusting tool (JDH-19) can be used to determine stroke control valve setting. Refer to text.

Valves located in housing (15—Fig. 176 or Fig. 177) control pump output as follows: The closed hydraulic system has no discharge except through the operating valves or components. Peak pressure is thus maintained for instant use. Pumping action is halted when line pressure reaches a given point by pressurizing the camshaft reservoir of pump housing, thereby holding pistons outward in their bores.

The cut-off point of pump is controlled by pressure of spring (27) and can be adjusted by turning adjusting screw (32). Adjustment procedure is given in paragraphs 153, 154 and 155. When pressure reaches the standby setting, valve (23) opens and meters the required amount of fluid at reduced pressure in crankcase section of pump. Crankcase outlet valve (18—Fig. 176 or 34—Fig. 177) is held closed by hydraulic pressure and blocks the outlet passage. When pressure drops as a result of system demands, crankcase outlet valve is opened by pressure of spring (19) and a temporary hydraulic balance on both ends of valve, dumping the pressurized crankcase fluid and pumping action resumes.

Cut-in pressure on late-style pump is determined by pressure of spring (19—Fig. 176). Spring (19) should test 63-77 N (14-17 lbs.) when compressed to a length of 74.5 mm (3 inches).

Cut-in pressure on early-style pump is determined by thickness of shim pack (43—Fig. 177) and pressure of spring (19). A special tool JDH-19B is available to determine thickness of shim pack. See Fig. 174. Assemble outlet valve (34—Fig. 177) and all components (19, 35, 36, 37, 38, 39, 40, 42, 44 and 45) using existing shim pack (43). Install special tool in place of plug (46) threading special tool into housing until gap (A—Fig. 174) is 3 mm (1/8 inch) as shown. If shim pack thickness is correct, scribe line on tool plunger should align with edge of tool plug bore (B). If not, remove stop plug (45) and add or remove shim washers (43) as required. Shims (43) are available in 0.25 and 0.76 mm (0.010 and 0.030 inch) thicknesses. If special tool JDH-19B is not available, use shim pack of same thickness as those removed, then add shims to raise cut-in pressure or remove shims to lower pressure.

On both early- and late-style pumps, add or remove shims (26—Fig. 175) as necessary to obtain pump shaft end play of 0.025-0.100 mm (0.001-0.004 inch). Renew all "O" rings, packings and seals and lubricate all parts with clean hydraulic oil. Install the stroke control housing (15—Fig. 176 or Fig. 177) on pump housing making sure that oil passages for both seal rings (7) are aligned. **Do not attempt to use an early-type housing (15—Fig. 177) with a late-type pump or a late-type housing (15—Fig. 176)**

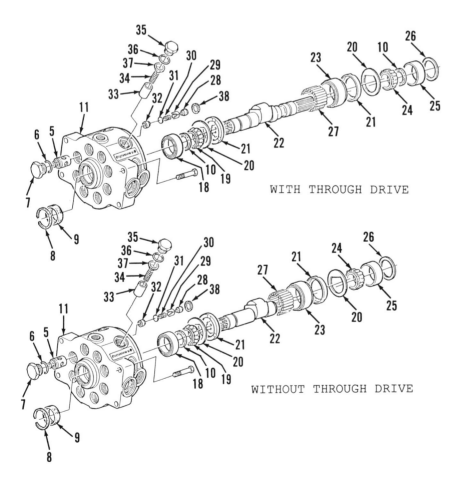

WITH THROUGH DRIVE

WITHOUT THROUGH DRIVE

Fig. 175—Exploded view of 40 cc (2.4 cubic inch) displacement hydraulic pumps used on some tractors.

5. Inlet valve (8)
6. "O" ring
7. Plug
8. Snap ring
9. Oil seal
10. Seal ring
11. Housing
18. Bearing cup
19. Bearing cone
20. Spacers
21. Thrust washers
22. Pump shaft (cam)
23. Race
24. Bearing cone
25. Bearing cup
26. Shim
27. Bearing needles (25)
28. Stop
29. Guide
30. Spring
31. Outlet valve (8)
32. Seat
33. Piston (8)
34. Spring
35. Plug
36. "O" ring
37. Sheath
38. Packing ring

with an early-type pump. Bolt holes will align, but high-pressure passages will NOT and damage or injury may result. Tighten stroke control valve housing (15—Fig. 177) retaining cap screws, plugs and other screws to the proper torque as follows:

Early pump (Without Serial No.)
Housing (15—Fig. 177) to body screws—

First torque .... 40-68 N•m
(30-50 ft.-lbs.)

Final torque .... 115 N•m
(85 ft.-lbs.)
Test port plug (17—Fig. 177) .... 34 N•m
(25 ft.-lbs.)
Adjusting screw lock nut (33—Fig. 177).... 60N•m
(45 ft.-lbs.)
Inlet valve plug (7—Fig. 175) .... 136 N•m
(100 ft.-lbs.)
Piston plugs (35—Fig. 175)............. 140 N•m
(100 ft.-lbs.)
Pump drive to pump shaft screws......... 50 N•m
(35 ft.-lbs.)
Pump mounting screws................. 120 N•m
(85 ft.-lbs.)

Late pump (With Serial No.)
Housing (15—Fig. 176) to body screws.... 120 N•m
(88 ft.-lbs.)

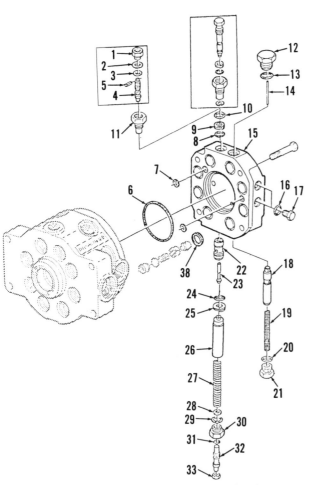

Fig. 176—Exploded view of stroke control valve assembly used on some 40 cc (2.4 cubic inch) displacement pump.

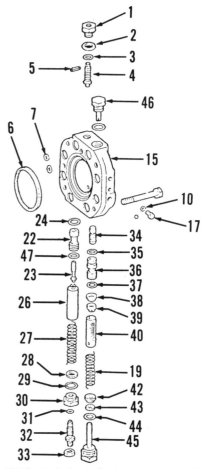

1. Bushing		
2. "O" ring	18. Outlet valve	
3. "O" ring	19. Spring	
4. Lock out screw	20. "O" ring	
5. Spring pin	21. Plug	
6. "O" ring	22. Sleeve	
7. Packing ring	23. Control valve	
8. "O" ring	24. "O" ring	
9. Washer	25. Back-up ring	
10. "O" ring	26. Guide	
11. Bushing	27. Spring	
12. Plug	28. Washer	
13. "O" ring	29. "O" ring	
14. Pin	30. Bushing	
15. Housing	31. "O" ring	
16. "O" ring (2)	32. Adjusting screw	
17. Plug (2)	33. Jam nut	

Fig. 177—Exploded view of pump cover and stroke control valve assembly used on some 40 cc (2.4 cubic inch) displacement pumps. Refer to text for differences between this type and that shown in Fig. 176. Refer to Fig. 176 for legend except the following.

34. Crankcase outlet valve	40. Filter
35. "O" ring	42. Seal (same as 38)
36. Crankcase outlet valve sleeve	43. Shims
37. "O" ring (same as 47)	44. "O" ring
38. Seal (same as 42)	45. Plug
39. Spring guide	46. Plug

Test port plug (17—Fig. 176) . . . . 34 N·m
                                           (25 ft.-lbs.)
Adjusting screw lock nut (33—Fig. 176) . . . 60 N·m
                                           (45 ft.-lbs.)
Piston plugs (35—Fig. 175). . . . . 185 N·m
                                           (135 ft.-lbs.)
Pump drive to pump shaft screws . . . . 50 N·m
                                           (35 ft.-lbs.)
Pump mounting screws. . . . . 120 N·m
                                           (85 ft.-lbs.)

On all models, reinstall pump by reversing the removal procedure and adjust standby pressure as outlined in paragraph 155.

ROCKSHAFT HOUSING, CYLINDER AND VALVE

All Models

166. REMOVE AND REINSTALL. On models without Sound Gard Body, disconnect battery ground straps, disconnect wires from starter safety switch and remove seat assembly. Disconnect lift links from rockshaft arms. Disconnect return line to rockshaft housing and lines from selective control valves to

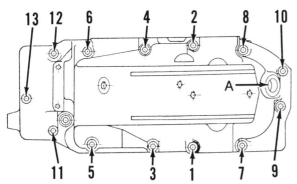

Fig. 178—Tighten screws attaching rockshaft housing to tractor in the sequence shown.

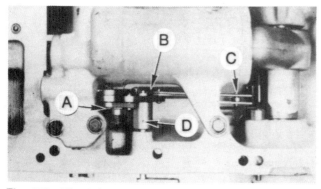

Fig. 179—View showing bottom side of typical rockshaft housing. Spring pin (D) retains control lever shaft to pivot block. Spring (B) is also shown at (9—Fig. 184).

remote couplers, then remove the breakaway couplers. Attach a hoist to rockshaft housing, place load selector lever in "MAX" position, then unbolt and lift off rockshaft housing assembly.

On models equipped with Sound Gard Body, remove Sound Gard Body as outlined in paragraph 174. Disconnect wires from starter safety switch. Disconnect lift links from rockshaft arms. Disconnect oil return line from selective control valves to rockshaft and lines to remote couplers. Move shift rod of load selector lever upward (MAX position). Attach a hoist to rockshaft housing, then unbolt and lift off rockshaft assembly.

When reinstalling, use a new gasket and reverse the removal procedure. Tighten mounting cap screws to a torque of 120 N·m (85 ft.-lbs.) in the sequence shown in Fig. 178.

167. OVERHAUL. With rockshaft housing assembly removed, disassemble as follows: If so equipped, unbolt and remove filler tube. Remove rockshaft arms from rockshaft. Remove selector lever linkage from load control arm. Remove adapter (6—Fig. 180) and the rate-of-drop adjusting parts (3, 4, 5, 6 and 7), then use a magnet to remove rate-of-drop ball (2) from hole. Remove front outside cylinder mounting cap screw, then turn unit so bottom side is accessible. Drive out spring pin (D—Fig. 179) that retains control shaft to pivot block. Remove snap ring (A) and unhook spring (B). Unbolt plate from rockshaft housing and remove quadrant and control shaft. Remove remote cylinder outlet adapter from left side of housing and unhook valve spring from linkage. Remove the four remaining cylinder attaching cap screws (33). Lift cylinder assembly from housing and disengage selector arm from roller link as cylinder is removed. If not previously removed, rate-of-drop ball (2) will fall.

Unbolt and remove cam (11—Fig. 182) from rockshaft, then pull rockshaft (4) from crank arm (5) and housing. Bushing (6—Fig. 181), "O" ring (5) and retainer (4) will be removed from the right side by the rockshaft as rockshaft is removed from right side. If necessary, separate piston rod (7—Fig. 182) from crank arm (5) by driving out spring pin (6). Load selector shaft can also be removed, if necessary. Remove flow control valve assembly from right front of hydraulic housing.

With components removed, disassemble cylinder and valve unit as follows: Remove plug (8—Fig. 180), spring (10) and check ball (2). Remove surge pressure relief valve (12). Remove retainer ring and pull linkage from linkage pivot pin. Remove snap rings (32), plugs (30), springs (27), sleeves (26), valve spools (25) and seats (24) from bores in housing. Keep valves identified so they can be reinstalled in original bores. Remove piston from cylinder by bumping open end of cylinder against a wood block or remove plug from front of cylinder and push piston out. If white seal ring (10—Fig. 182) is renewed with the black pressure ring (9), wait about 10 minutes until seal ring has contracted before attempting to install piston.

Clean and inspect all parts for undue wear or other damage. Check valve spring (10—Fig. 180) should test 18-21 N (4.0-4.7 lbs.) when compressed to a length of 19 mm (0.75 inch). Control valve springs (27) should test 43-51 N (10.0-11.5 lbs.) when compressed to a length of 23 mm (0.91 inch). Flow control valve spring (24—Fig. 181) should test 55-65 N (11.7-14.3 lbs.) when compressed to a length of 20 mm (0.79 inch). Inspect the seating area of the two control valves (25—Fig. 180) carefully. Leakage in this area could cause rockshaft settling if it occurs in the discharge (upper) valve, or upward rockshaft creep if it occurs in the pressure (lower) valve. Inspect seats (24) and sleeves (26) for any damage. Check rockshaft assembly for fractures, damaged splines or other damage. Check control linkage for worn or bent conditions. A worn adjusting cam (10—Fig. 184) will cause difficulty in adjusting lever neutral range.

When reassembling the rockshaft and cylinder unit, use sealant on threads of any pipe plugs that were removed and coat all "O" rings with oil. Start rockshaft into housing, align master splines of rockshaft (4—Fig. 182) and crank arm (5), then push rockshaft into position. Install rockshaft bushings (6—Fig. 181), which will be free in rockshaft bores, then install "O" rings (5) and retainers (4) with cupped side outward. Pre-assemble cam (11—Fig. 183), pin (14), snap rings (1) and tube (15) with offset hole in tube (15) for pin (14) positioned as shown. Install cam (11—Fig. 182) on rockshaft (4) and tighten screw (12) to 15 N·m (10 ft.-lbs.) torque. Attach connecting rod (7) to control arm (5). Insert valve seats (24—Fig. 180) in their original bores, small end first, then place valve spools (25) in seats with smaller diameter of spools in seats. Install sleeves (26), chamfered end first, and valve springs (27). Install "O" rings (28) and back-up rings (29) in bore of plugs (30) with back-up rings toward outer end. Install "O" rings (31) on outside of plugs, then carefully install plugs over ends of valves. Install back-up rings (15) and snap rings (32). If white seal ring (10—Fig. 182) is renewed with the black pressure ring (9), wait about 10 minutes until seal ring has contracted before attempting to install piston. Lubricate piston assembly and install it in cylinder. Refer to Fig. 180 and install check valve ball (2), spring (10) and plug (8) with new "O" ring (9). Install surge pressure relieve valve (12). Assemble linkage and install on linkage pivot pin. Reinstall flow control valve and, if not already done, remove rate-of-drop adjusting screw (7) from rockshaft housing. Install load selector control lever shaft (36—Fig. 181) in

Fig. 180—Exploded view of rockshaft valve components.

1. Cylinder housing
2. Ball
3. "O" ring
4. Washer
5. Snap ring
6. Adapter
7. Rate-of-drop screw
8. Plug
9. "O" ring
10. Spring
12. Surge relief valve
13. Cap screw
15. Back-up ring
24. Valve seat (2)
25. Valve spool (2)
26. Sleeve (2)
27. Spring (2)
28. "O" ring (2)
29. Back-up ring (2)
30. Plug (2)
31. "O" ring (2)
32. Snap ring (2)
33. Cap screws (4)
34. Plug
35. Pivot pin
36. Seal ring
37. Seal ring

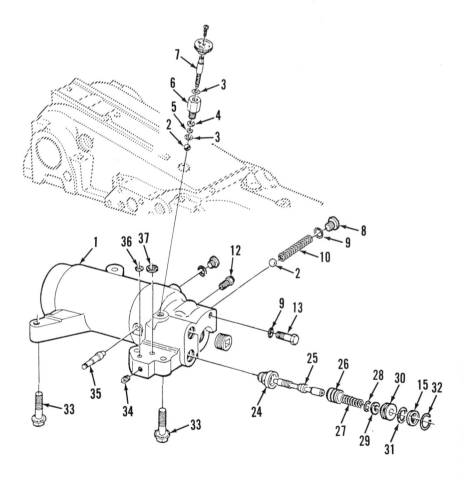

rockshaft housing. Install cylinder and valve assembly in rockshaft housing, while starting slot of roller link (2—Fig. 184) over selector control lever shaft, entering the connecting rod in piston and entering the rod (19—Fig. 185) into tube (15). Use new seal rings

and install front cap screw (13—Fig. 180) and the four screws (33). Install all five screws before first tightening front cap screw (13) to 70 N·m (50 ft.-lbs.), then tighten four remaining locking cap screws (33) to 70 N·m (50 ft.-lbs.) torque. Move the crank arm toward

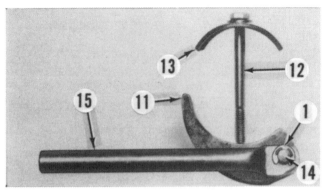

Fig. 183—View of tube (15) with offset hole for pin (14) correctly attached to cam (11).

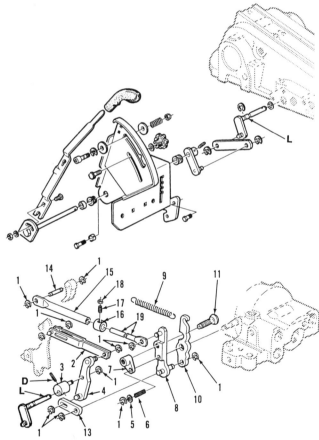

Fig. 181—Exploded view of typical rockshaft housing.

2.	Filler cap	27.	Plug
3.	Gasket	28.	Load selector lever
4.	Retainer	29.	Lever
5.	"O" ring	30.	Washer
6.	Bushing	31.	Nut
17.	Starter safety switch	32.	"O" ring
23.	Shim	33.	Bushing
24.	Spring	36.	Load control arm
25.	Flow control valve	37.	Gasket
26.	"O" ring	38.	Rockshaft housing

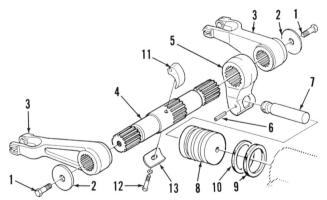

Fig. 182—View of rockshaft, piston and relative parts.

1.	Cap screw	8.	Piston
2.	Washer	9.	Pressure ring
3.	Lift arm	10.	Sealing ring
4.	Rockshaft	11.	Cam
5.	Crank arm	12.	Cap screw
6.	Spring pin	13.	Spacer
7.	Piston rod		

Fig. 184—Exploded view of control valve linkage. Spring (9) is also shown at "B" in Fig. 179.

D.	Spring pin	9.	Spring
L.	Lever	10.	Adjusting cam
1.	Retaining rings	11.	Adjusting screw
2.	Roller link	13.	Link
3.	Pivot block	14.	Pin
4.	Link	15.	Tube
5.	Washer	16.	Collar
6.	Spring	17.	Set screw
7.	Special nut	18.	Lock nut
8.	Link	19.	Rod

rear and hold a strip of metal 2-3 mm (0.08-0.12 inch) thick between crank arm (5—Fig. 185) and inside of rockshaft housing as shown. Loosen lock nut (18) and set screw (17), slide collar (16—Fig. 184) against tube (15), then tighten set screw and lock nut to retain position of the collar on rod. Install linkage (2, 3 and 4—Fig. 184), making sure that pin on link (4) engages link (13) and install snap ring (A—Fig. 179). Offset hole in pivot shaft (3—Fig. 184) should be down (with rockshaft inverted) when installing, then install pin (D). Connect the linkage spring and make sure that lever (L—Fig. 184) is correctly positioned. Turn rockshaft over to correct position, then install rate-of-drop ball (2—Fig. 180), adapter (6) and adjusting screw (7). Install control quadrant assembly and selector lever.

Reinstall rockshaft assembly by reversing the removal procedure and adjust as outlined in paragraphs 157 through 161.

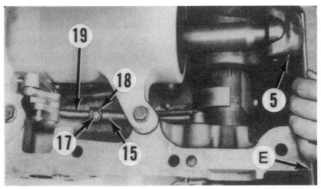

Fig. 185—View into rockshaft housing showing method of adjusting position of collar (16—Fig. 184). Spacer (E) is shown held in position between crank arm (5) and housing.

LOAD CONTROL (SENSING) SYSTEM

All Models

168. The load sensing mechanism is located in the rear of transmission case. Refer to Fig. 186 and Fig. 187 for views showing component parts.

Load control shaft (9) is mounted in tapered bushings. As load is applied to the shaft ends from the hitch links, the middle of shaft flexes forward and moves load control arm (20) that pivots on shaft (19). Movement of load control arm is transmitted to the rockshaft control valves via cam follower (27) and roller link (2—Fig. 184). The control linkage and control valves are opened or closed by this movement permitting oil to flow to or from the rockshaft cylinder.

169. R&R AND OVERHAUL. To remove the load sensing mechanism, remove rockshaft housing assembly as outlined in paragraph 166 and the three-point hitch. Remove cam follower spring (28—Fig. 187) and screw ¼ inch pipe plug, located near left final drive housing, out of transmission case. Slide pivot shaft (19) to the left out through pipe plug hole far enough so control arm (20) can be removed. Remove pin, retainer and bushing (4) from right end of load control shaft (9E or 9L) and bump shaft from transmission case. Negative stop screw (16) can also be removed if necessary.

Remove bearing housings with bushings (8). Renew bushings as necessary. New bushings are installed with larger chamfer out toward seal (5). Check special pin (15) for damage in area where it is contacted by load control shaft and renew if necessary. Also check contact areas of negative stop screw (16) or pins (6

Fig. 186—Drawing of typical load control mechanism and rockshaft control parts.

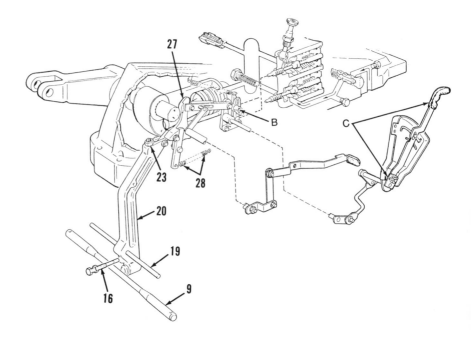

and 7) and load control arm (20). Check load control shaft (9E or 9L) to be sure it is not bent or otherwise damaged. Wear or damage to any other parts will be obvious.

Renew all "O" rings and seal rings and reassemble load sensing assembly by reversing disassembly pro-

cedure. When installing hitch draft links, check clearance between bushings and draft links. Add or remove shims at both sides to obtain a clearance of 2.5-3.7 mm (0.100-0.140 inch). After assembly is completed, adjust negative stop screw as outlined in paragraph 157.

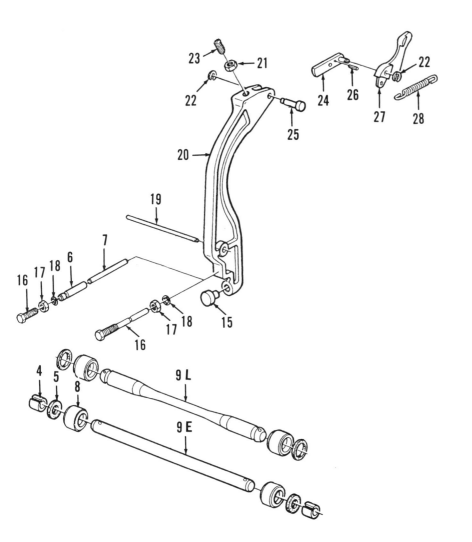

Fig. 187—Load control mechanism is located in rear of transmission case. Flexing of control shaft (9) actuates load control arm (20).

4. Bushing
5. Sealing washer
6. Rear pin (with IPTO)
7. Front pin (with IPTO)
8. Bushing
9E. Load control shaft (early)
9L. Load control shaft (later)
15. Special pin
16. Special screw
17. Jam nut
18. "O" ring
19. Control arm shaft
20. Load control arm
21. Jam nut
22. Retaining ring
23. Adjusting screw
24. Control arm extension
25. Pin
26. Pin
27. Cam follower
28. Spring

REMOTE CONTROL SYSTEM

Tractors may be equipped with single or dual selective (remote) control valves. Early models are equipped with poppet-type valves and later models are usually equipped with spool-type valves. Valves are mounted on a bracket attached to right final drive housing. Each control valve will operate a single- or double-acting remote cylinder.

SELECTIVE CONTROL VALVE

All Models (Spool Valve)

170. R&R AND OVERHAUL. To remove the valve or valves, disconnect hydraulic lines from

valves and, on models with Sound Gard Body, disconnect control lever linkage (Fig. 191). Remove cap screws (A—Fig. 189) and remove valves. Remove the four cap screws (B) and separate valves.

To remove valve cover and disassemble a valve set, drive pin (22—Fig. 188) from cover (21) and withdraw pinion shaft (25). Unbolt and remove cover (21), remove pins (19) and remove racks (20). Spool valve bodies (8) have hydraulic passages open to both sides. Valve spools and body are available only as a matched set and should never be interchanged with spool or body from another valve. One valve body (16) in a set has hydraulic openings open to only one side. Closed valve body (16) is assembled with three washers (9 and 18) that are identical. Single-acting selective control valves are similar to double-acting shown except only one valve spool and one rack is used.

Spring (12) should have a free length of 26 mm (1.02 inches) and should exert 210-260 N (49-58 lbs.) when compressed to 15 mm (0.59 inch). The area of valve spool at "V" seal (10) should be 9.96-10.00 mm (0.388-0.390 inch) and should not show evidence of wear. When assembling, sealing lips of seals (10) should be toward inside and the proper number of washers (9 and 18) should be installed. Tighten screws (24) to 15 N•m (11 ft.-lbs.) torque. Coat pinion shaft (25) with John Deere EP multipurpose grease, push valves into body until inner washer (11) is against shoulder of valve body and insert pinion shaft (25) so that bore of shaft is aligned with bore in cover for plug (23). Drive pin (22) into cover with slot away from pinion shaft. On models with Sound Gard Body, install lever (A—Fig. 190) pointing downward at 32 degrees as shown. On all models, tighten the four screws (B—Fig. 189) to 20 N•m (14 ft.-lbs.) and screws (A) to 50 N•m (35 ft.-lbs.) torque.

All Models (Poppet Valve)

171. R&R AND OVERHAUL. To remove the valve or valves, disconnect remote coupling lines, inlet line and return line from valve. On models with Sound Gard Body, disconnect control lever linkage (Fig. 191), unbolt valve from mounting brackets and remove valve. On models without Sound Gard Body, unbolt bracket from final drive housing and lift off valve and bracket assembly.

To overhaul the removed valve, refer to Fig. 192 and proceed as follows: Identify and remove control levers, then unbolt and remove end cap (37), which will contain metering valve (50). Note while loosening, that the four valve guides (33) are spring loaded and retained by the end cap. The four guides should move out with the cap as cap screws are loosened. If they do not, protect them from flying out as cap is removed. Withdraw guides (33) and springs (32 and 55), valves (31) and associated parts, keeping them identified and in proper order.

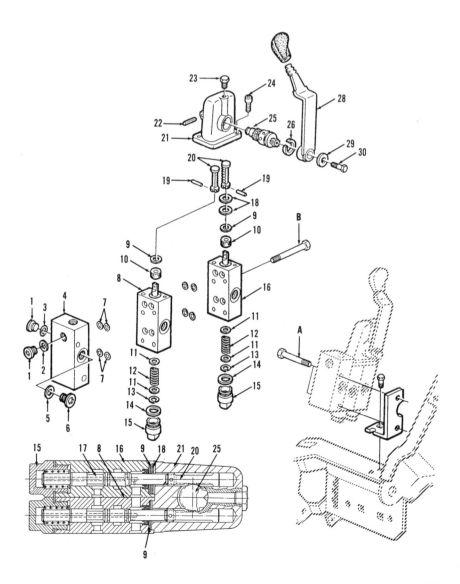

Fig. 188—Exploded and cross section views of spool-type selective (remote) control valve. Refer to Fig. 192 for poppet type.

1. Plug
2. Seal ring
3. Seal ring
4. Valve stack end
5. Seal ring
6. Plug
7. "O" rings
8. Spool & housing with through passages
9. Washer
10. Seal
11. Washer
12. Spring
13. Snap ring
14. Seal ring
15. End cap
16. Spool & end housing
18. Washers (same as 9)
19. Pins
20. Racks
21. Cover
22. Spring pin
23. Plug
24. Screws
25. Pinion shaft
26. Seal
28. Handle
29. Washer
30. Retaining screw

NOTE: Pressure springs are located on valves in location (P—Fig. 193) and should be color-coded red. Return springs coded green should be on valves located in ports (R).

Remove flow control valve (46—Fig. 192) and spring (47). Remove snap ring (26) and outer guide (23), then withdraw spring (36), detent cartridge assembly (18, 19, 20, 21, 22, 27 and 54) and detent follower assembly (15, 16 and 17).

Invert valve body, rocker assembly end up, then drive out spring pin (8) securing rocker (10) to lever shaft (60), withdraw shaft and lift out rocker assembly.

Clean all parts and inspect housing, side cover and end caps for cracks, nicks or burrs. Small imperfections can be removed with a fine file. However, parts should be renewed if their condition is questionable. Inspect poppet valves (31) and their seats in housing

(14) for grooves, scoring or excessive wear and renew parts as necessary. Check springs against the values that follow.

Flow control valve spring (47)—
  Free length . . . . 47 mm
                                        (1.85 inches)
  Test at 39 mm (1.5 inches). . . . . 160-195 N
                                        (36-44 lbs.)
Pressure valve springs (32 or 55)—
  Color . . . . Red
  Free length . . . . 44 mm
                                        (1.72 inches)
  Test at 31 mm(1.2 inches) . . . . 160-195 N
                                        (36-44 lbs
Return valve springs (32 or 55)—
  Color . . . . Green
  Free length . . . . 39 mm
                                        (1.5 inches)
  Test at 31 mm (1.2 inches). . . . . 80-100 N
                                        (18-22 lbs.)
Cam detent follower spring (36)—
  Free length . . . . 19 mm
                                        (0.75 inch)
  Test at 8 mm (0.31 inch) . . . . 58-70 N
                                        (13-16 lbs.)
Spring (11)—
  Free length . . . . 19.4 mm
                                        (0.76 inch)

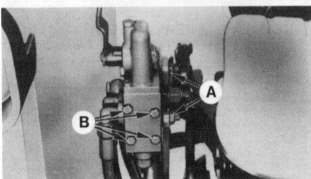
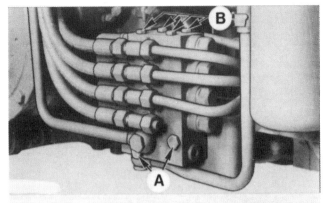

Fig. 189—Refer to text for proper removal and installation of selective control valves.

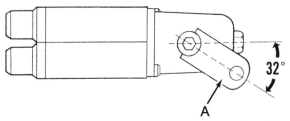

Fig. 190—Lever (A) should be installed pointing downward at 32 degree angle as shown for models with Sound Gard Body.

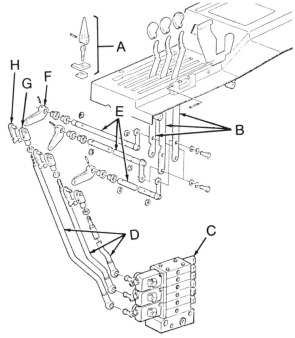

Fig. 191—Drawing of typical linkage connecting control handles to selective control valves on models with Sound Gard Body.

A. Stop (for hyd. motor operation)   E. Shafts
B. Control levers   F. Lever
C. Selective control valves   G. Fork end
D. Connecting rods   H. Safety strap

Test at 14 mm (0.56 inch) . . . . 89-108 N
(20-24.4 lbs.)
Spring (21)—
Free length . . . . 35 mm
(1.38 inches)
Test at 26 mm (1.02 inches) . . . . 51-62 N
(11.5-14 lbs.)

Wash housing with soap and water to remove any contamination, then dip all parts except keepers (6) in John Deere Hydraulic and Transmission oil before

assembling. Carefully install new "O" ring (24) and back-up ring (25) in housing bore, then install pre-assembled cartridge (18, 19, 20, 21, 22, 27 and 54), spring (36), guide (23) and snap ring (26). Install new "O" rings (61) in housing bores. Assemble rocker (5, 7, 9, 10, 11, 12 and 13) using Fig. 192 as a guide. The large float stop of detent cam (13) at arrow should be assembled opposite the larger pin boss (B). The regular cam (5) can be identified by chamfer (C) and different shape than float cam (11). Insert detent follower (15, 16 and 17) with roller (16—Fig. 194)

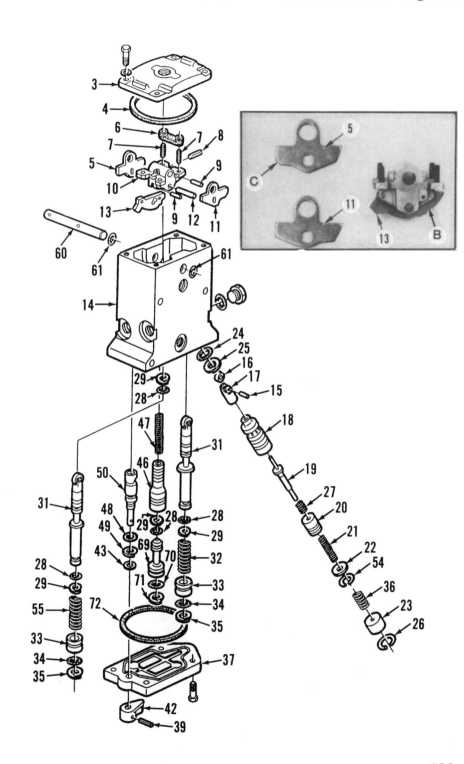

Fig. 192—Exploded view of typical poppet-type selective (remote) control valve. Refer to Fig. 188 for spool-type valve.

3. Cover
4. Packing
5. Regular cam
6. Rubber keeper
7. Adjusting screw
8. Spring pin
9. Drive pin
10. Rocker
11. Float cam
12. Pin
13. Detent cam
14. Housing
15. Spring pin
16. Roller
17. Detent follower
18. Detent cartridge
19. Detent pin
20. Detent piston
21. Spring
22. Special washer
23. Outer guide
24. "O" ring
25. Back-up ring
26. Snap ring
27. Spring
28. "O" ring
29. Back-up ring
31. Valves
32. Spring
33. Valve guide
34. "O" ring
35. Back-up ring
36. Spring
37. End cap
39. Pin
42. Metering lever
43. Thrust washer
46. Flow control valve
47. Spring
48. "O" ring
49. Back-up ring
50. Metering valve
54. Snap ring
55. Spring
60. Shaft
61. "O" ring
69. Flow control valve stop
70. "O" ring
71. Back-up ring
72. Packing

aligned with center line of housing, then install rocker assembly with detent cam (13) located as shown. Install control shaft (60) with lever positioned correctly, then install pin (8—Fig. 192) through rocker and shaft. Install new "O" rings (29) and back-up rings (28) on valves (31), then install valves (31) in proper bores as shown in Fig. 193. Adjust valve as follows before installing springs (32 or 55) and guides (33). Make sure roller on valve is properly aligned with detents (5 and 11—Fig. 192).

Selective control valve is adjustable and if disassembled, valve must be adjusted during assembly. Adjustment requires the use of a special adjusting cover JDH-15C and a dial indicator as shown in Fig.

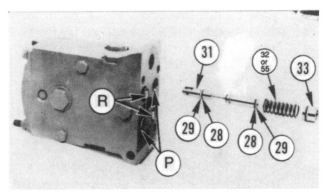

Fig. 193—Pressure valves are installed in bores (P) with RED springs and return valves are installed in bores (R) with GREEN springs.

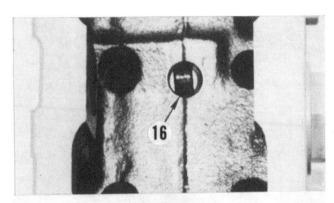

Fig. 194—View of selective control valve with rocker removed and with rocker (10) and shaft (60) being installed. Refer to text when assembling.

196. With valve guides (33—Fig. 192) removed, install adjusting cover (F—Fig. 196) with washers between housing and cover to ensure cover is level and secure.

NOTE: With selective control valves in neutral, pressure and return valves are adjusted by setting a specified distance between the operating cams and the valve rollers. This specified distance ensures that return valves open before the pressure valves and that maximum oil flow through selective control valves can be maintained.

Loosen all four screws (A, B, C and D—Fig. 195) to make sure there is some clearance. Finger tighten the four screws (G, H, J and K—Fig. 196) to seat the valves. Finger tighten detent screw (E) while moving control lever to locate neutral position. Install a dial indicator so that indicator pin contacts control lever 51 mm (2 inches) from centerline of control shaft and at right angles to lever. Dial indicator should have a minimum travel of 3.55 mm (0.140 inch) in each

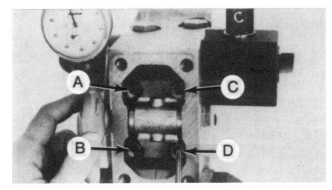

Fig. 195—View showing location of valve adjusting screws.

A. Pressure valve adjusting screw
B. Return valve adjusting screw
C. Return valve adjusting screw
D. Pressure valve adjusting screw

Fig. 196—Install dial indicator so indicator pin contacts control lever on a line 51 mm (2 inches) from centerline of control shaft. Install valve adjusting tool JDH-15C as shown.

E. Detent screw
F. Adjusting cover JDH-15C
G. Return screw
H. Pressure screw
J. Pressure screw
K. Return screw

direction from zero. Set indicator at zero which is detent cam zero. Turn each of the four adjusting screws (A, B, C and D—Fig. 195) in, one at a time, until dial indicator just begins to move, then back adjusting screws for pressure valves (A and D) out ½ turn and return adjusting screws (B and C) out ¼ turn.

Back out detent screw (E—Fig. 196) and screws (J and K) on one side. Check to make sure that screws (G and H) on other side are still tight. At opposite end of valve, adjust return screw (C—Fig. 195) so upward movement of control lever will read 0.5-0.8 mm (0.020-0.030 inch) on dial indicator. Turn screw IN to decrease dial indicator reading or OUT to increase. After adjusting screw (C), adjust pressure screw (D) so downward movement of control lever will read 1.5-1.8 mm (0.060-0.070 inch) on dial indicator. Then, adjust screws (A and B) of valves that are held against seats by screws (G and H—Fig. 196) as follows:

Tighten screws (J and K) and loosen screws (G and H). At opposite end of valve, adjust return screw (B—Fig. 195) until downward movement of control lever will read 0.5-0.8 mm (0.020-0.030 inch) from zero. Then, adjust pressure screw (A) until upward movement of control lever will read 1.5-1.8 mm (0.060-0.070 inch) from zero on dial indicator. Because this adjustment is so critical, it is advisable to adjust a second time using the same sequence.

With adjustment completed, remove adjustment cover (F—Fig. 196). Reassemble valve using new seals and gaskets. Springs color coded RED should be installed on pressure valves at locations (P—Fig. 193) and GREEN springs should be installed on return valves (R). Tighten control valve end cap and cover cap screws to a torque of 50 N·m (35 ft.-lbs.).

BREAKAWAY COUPLER

ISO Poppet Coupler

172. To disassemble the ISO poppet type remote coupler, remove plug (15—Fig. 197 or Fig. 198), screws (14 and 16), then remove dust cover assembly (5 through 8). Remove four screws (13) and separate

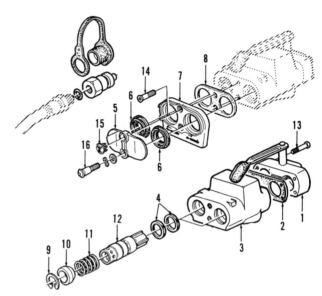

Fig. 197—Exploded view of one type breakaway coupler. Refer to Fig. 198 for other type.

1. Cover	9. Snap ring
2. Gasket	10. Sleeve
3. Housing	11. Spring
4. Seals	12. Receptacle
5. Cover	13. Screws
6. Grommets	14. Screw
7. Retainer	15. Plug
8. Gasket	16. Screw

Fig. 198—Exploded view of one type breakaway coupler. Refer to Fig. 197 for other type.

1. Lever & cover assy.	19. Snap ring
2. Gasket	20. Balls (10)
3. Housing	21. Coupler
4. Seal rings	22. Cone
5. Cover	23. Rubber washer
6. Grommet	24. Steel washer
7. Retainer	25. Spring
8. Gasket	26. "O" ring
9. Snap ring	27. Guide
10. Sleeve	28. "O" ring
11. Spring	29. Back-up ring
12. Receptacle	30. Piston
13. Screws	31. Back-up ring
14. Screw	32. "O" ring
15. Plug	33. Short spring
16. Screw	34. Ball
17. Back-up ring	35. Longer spring
18. "O" ring	36. Bleed valve
	37. Coupler socket

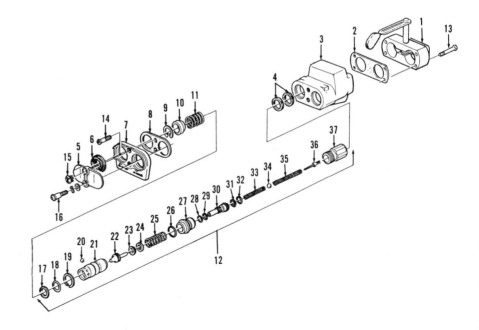

lever and cover assembly (1) from housing (3). Using special tool JDG266 or equivalent, depress sleeve (10) and remove snap ring (9), sleeve (10), spring (11) and receptacle (12). Remove the 10 balls from receptacle. Two types of receptacles are used.

To disassemble the coupler receptacle of the type shown in Fig. 198, clamp coupling (21) softly in a soft-jawed vise and unscrew socket (37). Disassembly, inspection and reassembly procedures will be evident. Always install new "O" rings, back-up rings and other soft parts when assembling. Retaining spring (11) should test 140-170 N (32-39 lbs.) when compressed to 19 mm (0.75 inch). Poppet valve spring (25) should test 75-90 N (16-20 lbs.) when compressed to 16.5 mm (0.64 inch). Bleed valve spring (35) should test 8-10 N (1.8-2.2 lbs.) when compressed to 1.65 mm (0.64 inch). Tighten coupler socket (37) to 50 N•m (35 ft.-lbs.) torque. Renew all "O" rings, back-up rings, seal rings and gaskets and reassemble by reversing the disassembly procedure. Tighten cap screws (13) retaining release lever and cover (1) to housing (3) to a torque of 8 N•m (6 ft.-lbs.). Tighten screws attaching coupler housing to bracket to 50 N•m (35 ft.-lbs.) torque. When installing breakaway coupler bracket to rockshaft housing, tighten cap screws to a torque of 120 N•m (85 ft.-lbs.).

To disassemble coupler receptacle of the type shown in Fig. 199, Fig. 200 and Fig. 201, proceed as follows. Remove snap ring (B—Fig. 199), bleed valve (C) and front piston (D). Remove snap ring (E), then withdraw rear piston (F), spring (G) and poppet (H). To disassemble bleed valve (C), refer to Fig. 200, remove snap ring (C), then remove spring (B) and bleed valve poppet (A).

Clean and inspect all parts. Spring (11—Fig. 197) should test 140-170 N (32-39 lbs.) when compressed to a length of 19 mm (0.75 inch). Spring (G—Fig. 199) should test 75-90 N (16-20 lbs.) when compressed to a length of 16.5 mm (0.64 inch). Spring (B—Fig. 200) should test 8-10 N (1.8-2.2 lbs.) when compressed to a length of 5 mm (0.20 inch). Renew all "O" rings, back-up rings, seal rings and gaskets and reassemble by reversing the disassembly procedure. Tighten cap

screws (13—Fig. 197) retaining release lever and cover (1) to housing (3) to a torque of 8 N•m (6 ft.-lbs.). When installing breakaway coupler on tractor, tighten cap screws to a torque of 120 N•m (85 ft.-lbs.).

REMOTE CYLINDER

All Models

173. To disassemble the double-acting remote cylinder, remove oil lines and end cap (18—Fig. 202). Remove stop valve (14) by pushing stop rod (9) completely into cylinder. Remove nut (21) from piston rod, then remove piston (24) and rod (34). Push stop rod (9) all the way into cylinder and drift out pin (27). Remove piston rod guide (26).

Renew all seals and examine other parts for wear or damage. Wiper seal (35) should be installed with lip toward outer end of bore. Install stop rod seal assembly (1, 2 and 3) with sealing edge toward cylinder. Complete the assembly by reversing the disassembly procedure. Tighten end cap screws to 160 N•m (120 ft.-lbs.) torque and piston rod guide screws to 50 N•m (35 ft.-lbs.) torque.

To adjust the working stroke, lift piston stop lever (29), slide adjustable stop (32) along piston rod to the desired position and press the stop lever down. If clamp does not hold securely, lift and rotate stop lever ½ turn clockwise and reset. Make certain that adjustable stop is located so that the stop rod contacts one of the flanges on adjustable stop.

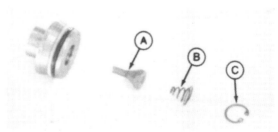

Fig. 200—Exploded view of bleed valve assembly (C—Fig. 199).

  A. Bleed valve
  B. Spring
  C. Snap ring

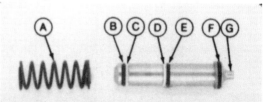

Fig. 201—View showing placement of "O" rings and back-up rings on poppet.

A. Spring	
B. "O" ring	E. "O" ring
C. Back-up ring	F. Packing
D. Back-up ring	G. Poppet

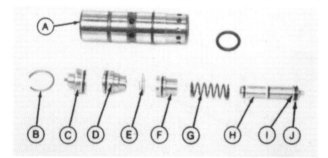

Fig. 199—Exploded view of ISO coupler receptacle.

A. Coupler receptacle	F. Rear piston
B. Snap ring	G. Spring
C. Bleed valve	H. Poppet
D. Front piston	I. Spacer
E. Snap ring	J. Packing

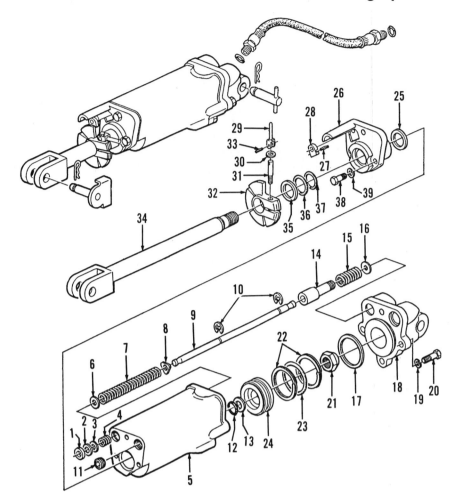

Fig. 202—Exploded view of typical remote cylinder.

1. Adapter	
2. Packing	21. Nut
3. Washer	22. Back-up ring
4. Spring	23. "O" ring
5. Cylinder	24. Piston
6. Washer	25. Gasket
7. Spring	26. Guide
8. Washer	27. Pin
9. Stop rod	28. Stop rod arm
10. Snap ring	29. Stop lever
11. Plug	30. Washer
12. "O" ring	31. Stop screw
13. Back-up ring	32. Rod stop
14. Stop valve	33. Pin
15. Spring	34. Piston rod
16. Gasket	35. Seal
17. Gasket	36. Washer
18. Cap	37. "O" ring
19. Lock washer	38. Cap screw
20. Cap screw	39. Lock washer

SOUND GARD BODY

Models So Equipped

174. REMOVE AND REINSTALL. To remove the Sound Gard Body, unbolt and remove left and right platform steps. Disconnect battery cables, then remove batteries and left hand battery box. Drain coolant, remove radiator side grilles, side panels and hood. Disconnect engine wiring harness at dash and disconnect steering hoses at connectors behind dash. Immediately plug line openings. Disconnect shut-off cable and speed control rod at fuel injection pump, then snap rear of speed control rod out of ball joint and remove rod. Remove rate-of-drop adjusting screw, left and right dash panels and the center platform plate. Drain systems, then disconnect brake and clutch hydraulic lines. Cover all openings to prevent the entrance of dirt into systems. Disconnect Hi-Lo shift rod, pto control rod, electrical connector plug and return line from top of transmission shift cover. Unplug rear electrical harness connector. If so equipped, disconnect air conditioning lines at couplers. Disconnect and lower the auxiliary fuel tank. Disconnect differential lock rod and after noting

length of hand brake rod, disconnect brake rod from rear yoke. Disconnect left and right brake hoses and cap or plug openings. Fully lower rockshaft lift arms, then detach lift link adjusters from cab frame. Remove rockshaft and selective control valve linkage rods. **Remove rods completely to prevent damage.** Loosen hose clamps and disconnect heater hoses. Disconnect transmission shift linkage and check for any hoses from cab that may be secured to parts that will remain with tractor. Unbolt Sound Gard Body supports, noting the arrangement of the support pads and washers.

Various methods of removing the Sound Gard Body may be used. When using a lifting bar, remove the two cap screws at top of Sound Gard Body and install two lifting eyebolts. Attach lifting crossbar and overhead hoist to the eyebolts. Carefully lift Sound Gard Body from tractor.

When installing Sound Gard Body, reverse removal procedure. Be sure rod and hoses are properly positioned before lowering body into position. See Fig. 203. Check for correct placement of support mounting pads and washers as shown in Fig. 204 or Fig. 205.

Tighten support nuts to a torque of 200 N·m (145 ft.-lbs.). Bleed steering, brake and clutch systems as described in appropriate paragraphs. Clean and apply refrigerant oil to air conditioning line connectors before attaching, then check air conditioning charge as required.

Fig. 203—Check to be sure that all control rods (A), hoses and connectors (B) are correctly positioned before lowering cab onto tractor.

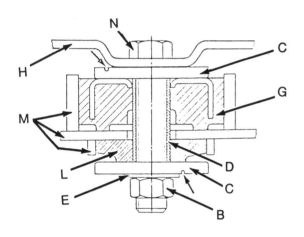

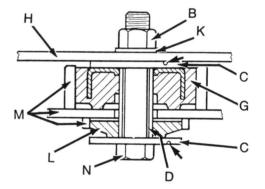

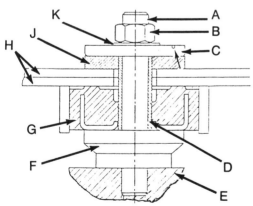

Fig. 204—Sound Gard Body mounts for all models except 2955 should be assembled as shown. Marked side of washers (C) indicated by arrows should be away from rubber parts. Refer to Fig. 205 for 2955 models.

A. Stud
B. Nut
C. Washers
D. Bushing
E. Final drive housing
F. Washer
G. Rubber bearing pad
H. Body frame
J. Rubber washer
K. Washer
L. Rubber washer
M. Mounting bracket
N. Cap screw

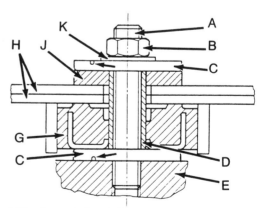

Fig. 205—Sound Gard Body mounts for 2955 models should be assembled as shown. Marked side of washers (C) indicated by arrows should be away from rubber parts. Refer to Fig. 204 for other models.

A. Stud
B. Nut
C. Washers
D. Bushing
E. Final drive housing
G. Rubber bearing pad
H. Body frame
J. Rubber washer
K. Washer
L. Rubber washer
M. Mounting bracket
N. Cap screw

NOTES

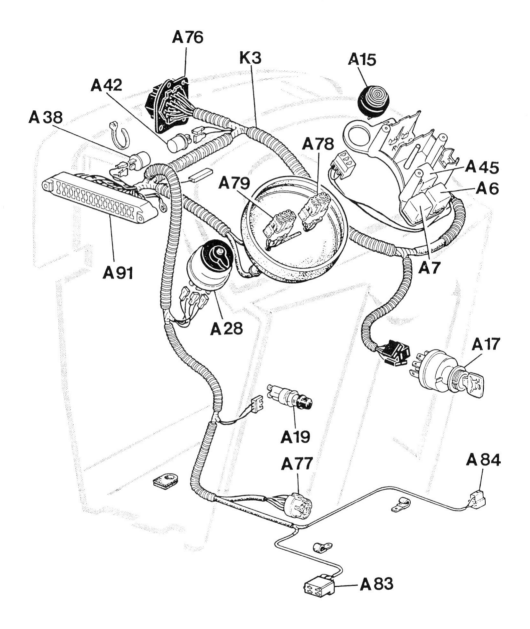

Fig. W1—Dash wiring harness typical of models before serial number 637 100L without Sound Gard body or Roll Over Protection.

Legend for wiring diagrams Fig. W1, Fig. W2 and Fig. W3.

A1. Alternator
A2. Starting motor
A3. Batteries
A4. Alternator indicator light
A5. Fuse for neutral start switch (0.5 amp.)
A6. Starter relay
A7. Relay for handbrake indicator light
A8. Neutral start switch
A10. Sending unit for handbrake indicator light
A11. Diode
A12. Diode
A13. Warning light
A14. Fuse 6
A15. Acoustical handbrake warning signal
A16. Fuse 11
A17. Main switch
A18. Fuse 4
A19. Button for Ether or electric starting aid
A20. Relay for electric starting aid
A21. Glow plug for electric starting aid
A22. Solenoid for Ether or electric starting aid
A23. Fuel preheater switch
A24. Thermostat switch
A25. Relay for preheater
A26. Glow plug for fuel preheater
A27. Fuse 9
A28. Light switch
A29. Fuse 7
A30. Fuse 2
A31. Fuse 1
A32. Fuse 3
A33. Lightbar
A34. Worklight & tail light
A35. Outlet socket
A36. Left hand headlight
A37. Right hand headlight
A38. Flasher
A39. Left turn signal indicator light
A40. Right turn signal indicator light
A41. Turn signal indicator warning light
A42. Flasher
A43. Sending unit for engine oil pressure
A44. Transmission oil pressure warning switch
A45. Time delay relay
A46. Engine oil pressure warning light
A47. Transmission oil pressure warning light
A48. Handbrake indicator light
A49. Air cleaner restriction indicator light
A50. Air cleaner restriction warning switch
A51. Hourmeter
A52. Diode

A53. Toggle switch for front-wheel drive
A54. Front-wheel drive indicator light
A55. Solenoid for front-wheel drive
A56. Diode
A57. Coolant temperature gage
A58. Fuel gage
A59. Coolant temperature sending unit
A60. Fuel gage sending unit
A61. Diode
A62. Diode for air cleaner restriction indicator light
A63. Diode for engine oil pressure warning light
A64. Diode for transmission oil pressure warning light
A65. Fuse 13
A66. Relay for front pto
A67. Front pto indicator light
A68. Switch for front pto
A69. Solenoid for front pto
A70. Diode
A71. Fuse 14
A72. Horn button
A73. Horn
A74. Cigarette lighter
A75. Diode
A76. Engine connector (21-pin)
A77. Transmission connector (12-pin)
A78. Right hand instrument cluster connector (12-pin)
A79. Left hand instrument cluster connector (12-pin)
A80. Fuse 8
A81. Diode
A82. Fuse (20 amp.)
A83. Left hand fender connector (4-pin)
A84. Right hand fender connector (2-pin)
A85. Left hand instrument cluster light
A86. Right hand instrument cluster light
A87. Diode
A88. Diode
A89. Engine & tractor speed meter
A90. Magnetic sending unit for engine & tractor speed meter
A91. Fuse box
A92. Rear lightbar connector (2-pin)
A93. Front lightbar connector (2-pin)
A94. Right fender connector (3-pin)
A95. Left fender connector (7-pin)
A96. Starter relay
A97. Fuse 12
A98. Fuse 16
A99. Connector
A100. High beam indicator light
A101. Fuse 15
A102. Diode

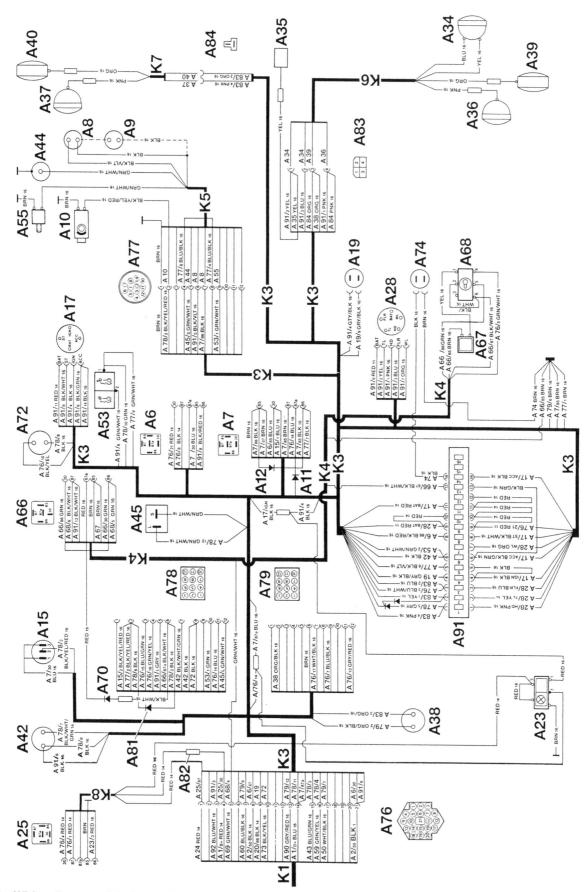

Fig. W2—Wiring diagram of dash panel, transmission and fenders typical of models before serial number 637 100L without Sound Gard Body or Roll Over Protection.

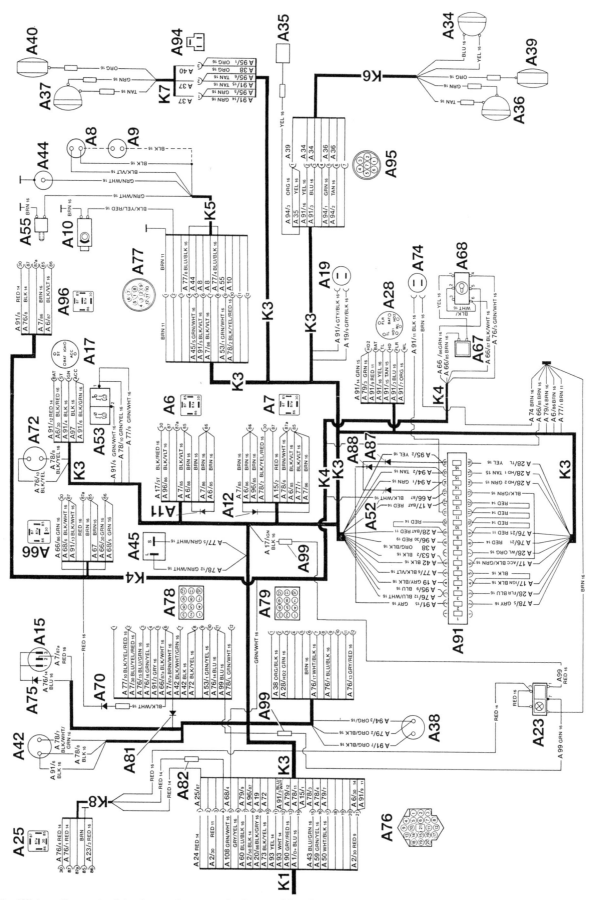

Fig. W3—Wiring diagram of dash panel, transmission and fenders typical of models beginning at serial number 637 100L without Sound Gard Body or Roll Over Protection.

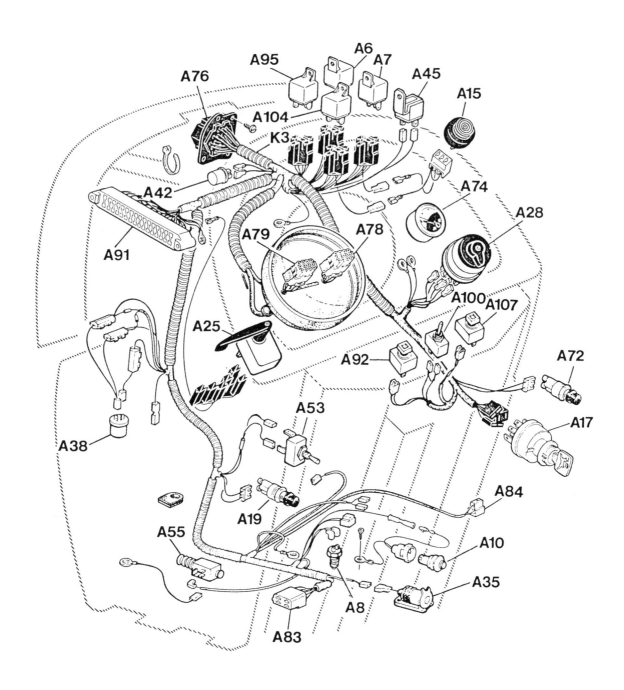

Fig. W4—Dash wiring harness typical of GP models before serial number 671 800L without Sound Gard Body or Roll Over Protection.

Legend for wiring diagrams Fig. W4 and Fig. W5.

A95. Relay for acoustical handbrake warning signal
A96. Fuse 16
A97. Fuse 17
A98. Diode
A99. Diode
A100. Backup alarm switch
A101. Sending unit for backup alarm (w/reverser)
A102. Sending unit for backup alarm (w/o reverser)
A103. Horn for backup alarm
A104. Coolant temperature buzz light relay
A105. Fuse 10
A106. Coolant temperature buzz light sending unit
A107. Coolant temperature buzz light
A108. Front pto connector (2-pin)
A109. High beam indicator light
A110. Engine oil pressure gage sending unit
A111. Diode

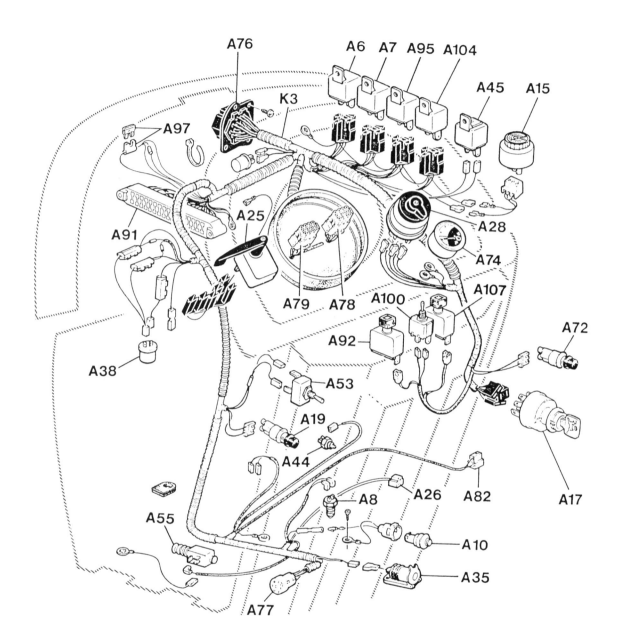

Fig. W5—Dash wiring harness typical of GP models beginning at serial number 671 800L without Sound Gard Body or Roll Over Protection.

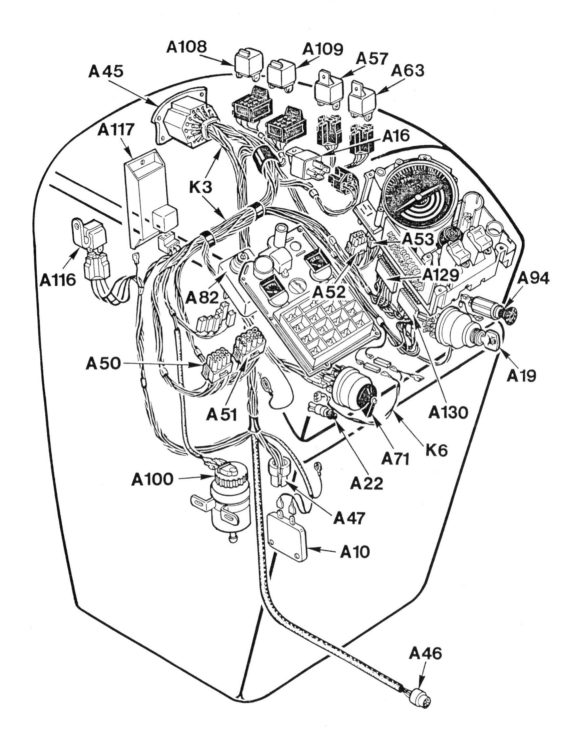

Fig. W6—Dash wiring harness typical of models before serial number 645 249L with Sound Gard Body or Roll Over Protection.

Legend for wiring diagrams Fig. W6 and Fig. W7.

A10.   Handbrake indicator light sending unit
A16.   Starter relay
A19.   Main switch
A22.   Ether or electric starting aid button
A45.   Engine connector (21-pin)
A46.   Transmission connector (12-pin)
A47.   Cab connector (19-pin)
A50.   Left hand dash panel connector (12-pin)
A51.   Left hand dash panel connector (12-pin)
A52.   Right hand dash panel connector (12-pin)
A53.   Right hand dash panel connector (12-pin)
A57.   Front work light relay
A63.   Rear work light relay
A71.   Light switch
A82.   Turn signal indicator light switch
A94.   Socket for hand lamp
A100. Clutch indicator light sending unit
A108. Relay for 7 terminal socket
A109. Relay for 7 terminal socket
A116. Relay for front wheel drive indicator light
A117. Not used
A129. Printed circuit board fuse plug (9-pin)
A130. Printed circuit board fuse plug (9-pin)

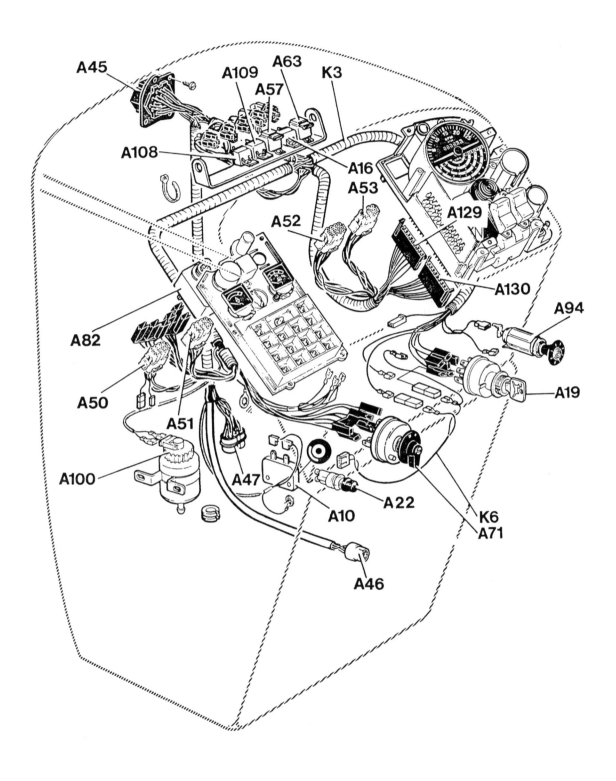

Fig. W7—Dash wiring harness typical of models after serial number 645 248L with Sound Gard Body or Roll Over Protection.

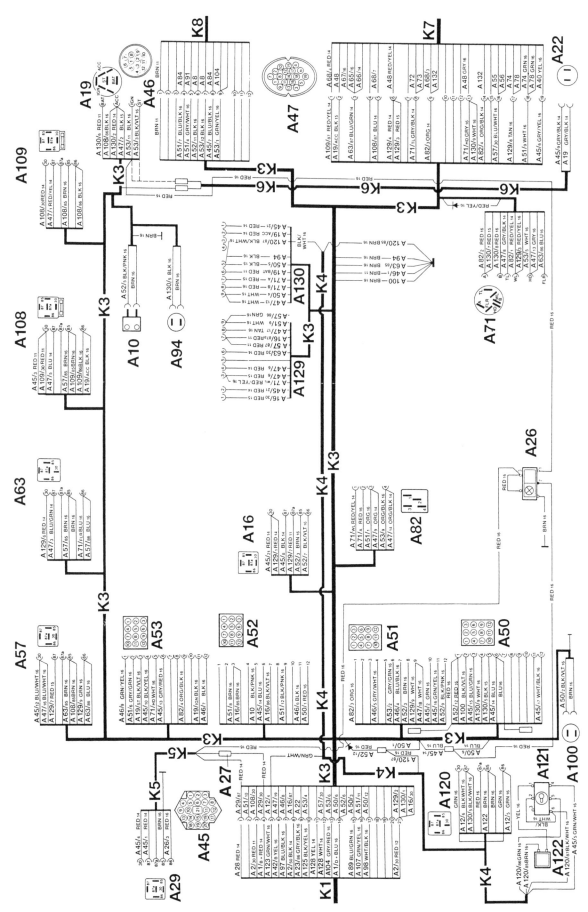

Fig. W8—Wiring diagram of dash panel typical of models with Sound Gard Body or Roll Over Protection.

Legend for wiring diagrams Fig. W8, Fig. W9 and Fig. W10.

A8. Neutral start switch
A33. Shift console light
A34. Cab interior light
A36. Cab interior light door switch
A37. Cab relay
A38. Blower fan
A40. Thermostat switch
A46. Transmission connector (12-pin)
A55. Left front work light
A64. Left rear work light
A65. Right rear work light
A68. Socket (7 terminal)
A72. Left tail light
A73. Right tail light
A74. Left headlight
A76. Front turn signal & warning lights
A77. Rear turn signal & warning lights
A79. Front turn signal & warning lights
A80. Rear turn signal & warning lights
A84. Auxiliary tank fuel gage sending unit
A91. Transmission oil pressure warning switch
A96. Diode
A104. Front wheel drive solenoid
A108. Relay for 7 terminal socket
A109. Relay for 7 terminal socket
A110. Left hand windshield wiper
A111. Right hand windshield wiper
A112. Left hand windshield wiper switch
A113. Right hand windshield wiper switch
A119. Not used on these models
A129. Printed circuit board fuse (9-pin)
A130. Printed circuit board fuse (9-pin)
A150. Ammeter
A152. Voltmeter
A153. Coolant temperature buzz light
A154. Engine oil pressure buzz light
A156. Engine oil pressure buzz light relay
A157. High pressure switch (some late models)
A158. Low pressure switch (some late models)
A159. Diode (some late models)

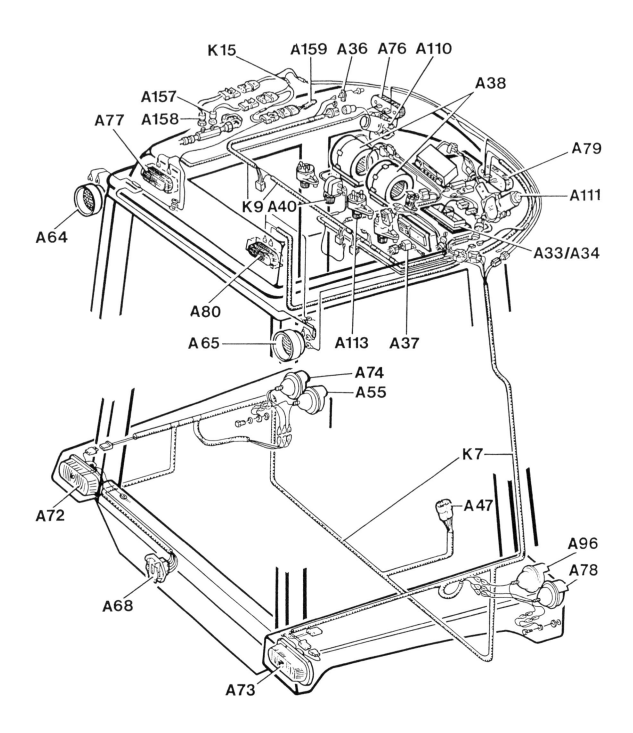

Fig. W9—Sound Gard Body wiring harness typical of models so equipped.

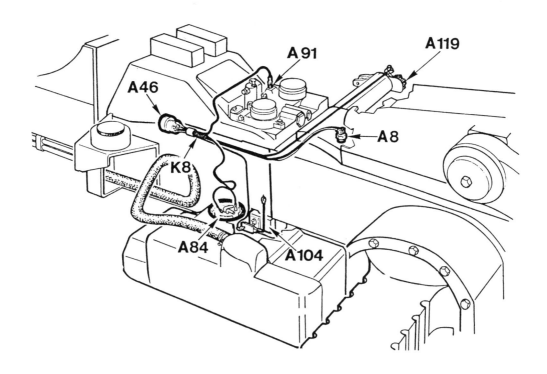

A91

A119

A46

A8

K8

A84

A104

Fig. W10—Transmission wiring harness typical of models with Sound Gard Body.

NOTES

NOTES

NOTES

NOTES

Technical Information

Technical information is available from John Deere. Some of this information is available in electronic as well as printed form. Order from your John Deere dealer or call **1-800-522-7448**. Please have available the model number, serial number, and name of the product.

Available information includes:

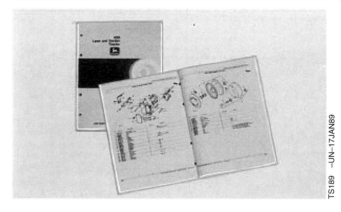

- PARTS CATALOGS list service parts available for your machine with exploded view illustrations to help you identify the correct parts. It is also useful in assembling and disassembling.
- OPERATOR'S MANUALS providing safety, operating, maintenance, and service information. These manuals and safety signs on your machine may also be available in other languages.
- OPERATOR'S VIDEO TAPES showing highlights of safety, operating, maintenance, and service information. These tapes may be available in multiple languages and formats.

- TECHNICAL MANUALS outlining service information for your machine. Included are specifications, illustrated assembly and disassembly procedures, hydraulic oil flow diagrams, and wiring diagrams. Some products have separate manuals for repair and diagnostic information. Some components, such as engines, are available in separate component technical manuals
- FUNDAMENTAL MANUALS detailing basic information regardless of manufacturer:
  - Agricultural Primer series covers technology in farming and ranching, featuring subjects like computers, the Internet, and precision farming.
  - Farm Business Management series examines "real-world" problems and offers practical solutions in the areas of marketing, financing, equipment selection, and compliance.
  - Fundamentals of Services manuals show you how to repair and maintain off-road equipment.
  - Fundamentals of Machine Operation manuals explain machine capacities and adjustments, how to improve machine performance, and how to eliminate unnecessary field operations.

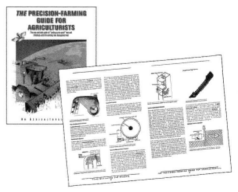

TS189 –UN–17JAN89

TS191 –UN–02DEC88

TS224 –UN–17JAN89

TS1663 –UN–10OCT97